TPM for Operators

Compiled by Productivity Press

Kunio Shirose, Advisory Editor

Publisher's Message by
Norman Bodek

Productivity Press

Portland, Oregon

Copies of the books from which this guide was compiled, or additional copies of this guide are available from the publisher. Address all inquiries to:

Productivity Press
P.O. Box 13390
Portland, OR 97213-0390
Telephone: (503) 235-0600
Telefax: (503) 235-0909

Cover design by Gary Ragaglia
Printed and bound by Goodway Graphics
Printed in the United States of America on acid-free paper

Library of Congress Cataloging-in-Publication data

TPM for operators/compiled by Productivity Press; Kunio Shirose, advisory editor.
p. cm.
ISBN 1-56327-016-1
1. Plant maintenance. I. Shirose, Kunio. II. Productivity Press (Portland, OR)
TS192.T69 1992
658.2'7—dc20 92-21526
CIP

97 96 95 94 10 9 8 7 6 5 4

Contents

Publisher's Message

This book is intended to give you, the equipment operator, the information you need to understand total productive maintenance (TPM) and your role in it.

The first three chapters describe what TPM is, why it is important, how it developed, and what its goals and characteristics are. The last three chapters tell you about the TPM approach to making improvements in your equipment and your workplace. You will also find an outline of an overall company plan for TPM and a description of your role, as operator, in making TPM work.

Since every plant is unique, TPM cannot be put into practice in exactly the same way everywhere. However, success always depends on the participation and cooperation of all employees. From top management to shop floor, everyone must do his or her part.

To remain competitive in today's global marketplace, a company needs a way to maintain its production machinery in top condition. TPM is the way to do it. TPM can form the foundation for improvements to the entire production process. It has been defined as a set of activities for restoring equipment to its optimal conditions and changing the work environment to maintain those conditions.

This definition of TPM is deceptively simple. Maintaining optimal equipment conditions means more than just making sure each machine runs well. It means ensuring that it runs so well that it never breaks down; always operates at designed speed or faster with no idling or minor stoppages; never produces a defective product; and causes a minimum of start-up, setup, and adjustment losses. Besides this, it means establishing and maintaining standardized methods for equipment diagnosis, early detection of abnormalities, spare parts management, parts replacement procedures, and information systems to record equipment histories and breakdown data. An effective TPM program also aims at minimizing the impact of equipment deterioration and establishes a method to help design engineers incorporate improvements into new equipment.

TPM is not simply a maintenance program. It requires the cooperation and involvement of all levels and divisions of the company, the breaking down of traditional attitudes toward specialization, and the establishment of educational systems designed to upgrade skill levels of maintenance and production personnel.

In a company implementing TPM, production employees are organized into groups or teams, each responsible for learning to care for their equipment and perform various maintenance tasks. This book provides an overview of the philosophy of TPM and describes activities for each step toward optimizing equipment effectiveness and establishing a shop-floor maintenance system.

TPM for Operators is the first in a series of books Productivity Press plans to produce to serve the information needs of front-line workers. If there are particular topics you would like to see included in short guides such as this one, please write to us at the address on the back of the title page.

Norman Bodek
Publisher
Productivity Press

Preface

I am pleased to participate in the publication of *TPM for Operators*. I believe it will be a helpful reference for everyone involved in teaching, learning, and promoting TPM.

TPM activities produce good results and contribute to manufacturing cost reduction. *Maintenance* means keeping equipment and its operators in perfect condition. This translates into:

- building equipment that does not break down, produce defects, or have minor stoppages
- training operators who are capable of detecting defects, implementing inspection and lubrication for maintenance, and exchanging parts easily
- creating a worksite where machines are simple to operate, work can be done easily and safely, and productivity is high.

The ultimate goal of TPM is to implement "perfect manufacturing." To achieve this goal, autonomous maintenance activities teach operators how to:

- detect defects
- make continuous improvements
- find satisfaction in improvement activities
- set operating standards
- understand equipment mechanisms.

Moreover, operators must be capable of implementing these items.

Production technology and maintenance staff also need to understand the basic concepts of chronic loss (breakdown, defects, and small line stops) and zero defects, accidents, and so on, so that they can implement improvements in these areas.

TPM activities did not have a smooth start in Japan; it was a difficult experience of trial and error. The reasons for failure included:

- Insufficient understanding of basic concepts of TPM
- Lack of qualified people who can teach TPM
- Lack of understanding of the depth to which TPM should be implemented in different areas
- Poor teaching of concrete methodology.

This book provides simple explanations of basic TPM concepts to help you, the American equipment operator, understand TPM correctly. I encourage you to look beyond differences of language, culture, and ways of thinking to learn the power of TPM. By reading and rereading this book, all employees can gain a better understanding of TPM.

TPM has been implemented not only in Japanese industries but also in South Korea, Indonesia, Singapore, and

Malaysia. I would like to encourage the spread of TPM implementation in the United States and welcome feedback from American employees on how this book can better serve that purpose.

Kunio Shirose

1
Causes of Breakdowns and Defects

Chronic breakdowns and defects have many causes, one of which is people, as the following example shows.

> A worker reports an equipment breakdown. The production supervisor is very concerned about meeting delivery schedules, but the production line obviously had to be stopped to deal with the problem. A maintenance worker is called in and, after some troubleshooting, reports that the damage is serious enough that certain parts need to be replaced. The machine could be down as much as four or five hours.
>
> As a temporary measure, the production supervisor decides to shift production to another machine while the repairs are made. At this point, the maintenance worker says, "You know, this breakdown couldn't just happen out of the blue. I'll bet this machine's been acting funny for at least two or three days, and this part was probably vibrating or making a strange sound. You guys should have told me about this sooner. I could have fixed it in about ten minutes during your lunch break. Please let me know as soon as you notice anything strange about any of these machines, OK?"

The supervisor agreed, but said the maintenance department should teach the operators how to recognize the early signs of these problems. This seemed like a good idea, so they decided to set a meeting time.

After thinking a moment, the supervisor realized that this was not a long-term solution either and said, "But why does this kind of breakdown happen? Sometimes certain parts are *always* breaking and other times, like today, some other part suddenly breaks. We don't know if we're using the equipment wrong, or if the equipment itself is bad, or if the maintenance procedures are no good. Whatever the cause, it sure fouls up our work!"

There is nothing unusual about this situation. In fact, most workplaces suffer from the same dilemma. Breakdowns are the root of all problems, because when they occur, production stops, deliveries are delayed, and product defects are created; in other words, a single breakdown can wreak havoc throughout the factory and can "break down" the entire operation.

In this situation, the damaged equipment probably signaled its abnormal condition through unusual vibrations or noises. If the workers had read these signs and responded promptly, the breakdown could have been avoided; so the human factor was the root cause of the breakdown.

Whenever you look deeply enough into the reasons for *any* breakdown, you always find people-related causes. So if people are ultimately responsible for creating breakdowns, why shouldn't they also be able to prevent them? There are ways to reduce and even completely eliminate breakdowns, and one is the subject of this book: TPM (total productive maintenance).

A detailed definition will be provided later. For now, understand it as something that involves the cooperation of maintenance staff, who act as "equipment doctors," and production staff, who work with the equipment every day.

Equipment operators need to learn how to discover abnormalities so they can identify and report problems early on.

IT'S A WHAT, NOT AGAIN!! GOOD THING YOU STOPPED THE LINE!! NOW GET SOMEONE FROM MAINTENANCE!
RIGHT !!
HOW LONG WILL IT TAKE?
IF WE DON'T HAVE THE PARTS, FOUR OR FIVE HOURS.
BUT WE'LL MISS THE DELIVERY DEADLINE!!
LISTEN, YOU'VE GOT TO TELL US WHEN THE MACHINES ARE ACTING STRANGELY. DON'T LET IT GO SO LONG!!
BUT YOU DON'T TELL US WHAT TO LOOK FOR!
?

Maintenance staff must analyze equipment data provided by the production department, find the causes of breakdowns and abnormalities, and take effective measures to prevent similar problems from happening again.

The road to achieving zero breakdowns starts with daily equipment checks and other activities performed by equipment operators, as well as specialized activities performed by maintenance workers.

What conditions *promote* breakdowns and defects in typical workshop environments? To answer this, look at the overall workshop conditions, which express the basic attitudes of the people who use the workshop. Consider which of the following conditions exist in your own workplace:

Equipment conditions

- The equipment is generally very dirty.
- Cutting debris is scattered on and around the equipment.
- The equipment leaks hydraulic fluid and lubricants.
- Oil pans are overflowing.
- People don't mind seeing dirt and grime piling up everywhere: they accept it as normal.
- Motors are coated with a layer of oil mist.
- Motors get very hot or make strange noises.
- Switches are covered with oil.
- Large covers are sometimes used to protect certain machines, but their internal parts are not cleaned.
- Some parts rattle and vibrate.
- The equipment is positioned to make access for routine maintenance checks difficult.
- Oil cans are left empty and dirty.
- Drains are clogged.
- Wires and pipes are left in chaotic configurations, making it hard to see which one goes where.

Area around equipment

- It takes a lot of time to clean up.
- The floor is left dirty and, in spots, slippery with oil.
- Dies and fixtures are disorganized and dirty.
- There are a lot of useless items lying around.
- Things are not kept in specified places.

Equipment operators

- Operators do not perform regular equipment checks. In fact, they do not even know *how*.
- When equipment must be oiled, only some operators know when and where to oil, and how much to use. Even those who know the oiling schedule and procedure do not always perform it correctly.
- Operators do not know how to replace equipment parts or perform precision checks.
- When operators find an abnormality, they call the maintenance staff without trying to understand the problem.
- Operators do not regard breakdowns and defects as their own problems.

Equipment losses

- Equipment breakdowns occur frequently, at a rate of 3 percent or higher.
- It takes a long time to fix minor problems, and often the repair is only temporary.
- Problems following changeover occur more (or less) often, depending on who does the changeover.
- Changeover and adjustments take a lot of time, and people accept adjustments as natural.

- Idling and minor stoppages happen often, preventing automated processes from operating for long periods unassisted.
- Rework is required at a rate of 3 percent and has become chronic.
- It is very hard to understand the causes of problems that make rework necessary.
- Processing speed has been decreased because too many defects occurred at the rated speed.

If any of these conditions exist in your workshop, it is time for a change. Done correctly, TPM will get rid of these conditions and ensure high productivity and quality (see Figure 1). Like the building of Rome, however, TPM cannot be fully implemented in a day. As implied by the *total* in total productive maintenance, effective TPM requires the participation of everyone — from management to floor workers.

Number of breakdowns
Company A: Reduced from 4,800 to 110 (1/40 of previous level)
Company B: Reduced from 860 to 6 (1/100 of previous level)

Defect /Repair Rate
Company A: Reduced from 3.0% to 0.1% (1/30 of previous level)
Company B: Reduced from 2.0% to 0.05% (1/40 of previous level)

Cleaning Time
Company A: Reduced from 40 minutes to 10 minutes (1/4 of previous level)
Company B: Reduced from 45 minutes to 7 minutes (1/6 of previous level)

Volume of Hydraulic Fluid and Lubricant Supplies
Company A: Reduced from 3,000 liters to 1,000 liters (1/3 of previous level)
Company B: Reduced from 4,000 liters to 800 liters (1/5 of previous level)

Figure 1. TPM Successes

2

What Is Total Productive Maintenance?

TPM is a new approach to equipment management. *Equipment management* is the set of activities that prevents quality defects and breakdowns, eliminates the need for equipment adjustments, and makes the work easier and safer for equipment operators.

How did TPM evolve?

HISTORY OF EQUIPMENT MANAGEMENT

The concept of PM (preventive maintenance) was introduced in 1951. PM is a program of planned inspections, replacements, and repairs designed to prevent expensive catastrophic failures and control deterioration. Before PM, companies generally practiced BM (breakdown maintenance), which means fixing equipment only after it has broken down. Companies that adopted the concept of preventive maintenance greatly reduced equipment breakdowns.

Over the years the PM approach gradually changed to meet the new demands placed on industry. One such change was the introduction of the concept of MI (maintainability and methods improvement), which goes beyond the restorative type of repairs performed as part of preventive maintenance. MI promotes modifications or procedures that make the same breakdown less likely to happen again. Another change came with the concept of MP (maintenance prevention), which involves equipment designers in building better equipment that is easier to maintain.

Finally, the PM, MI, and MP approaches were consolidated under a new definition of PM, which in this case stands for productive maintenance. Productive maintenance is aimed toward maximizing productivity — which means profitability. To achieve this goal, it includes four types of activities:

- Preventive maintenance
- Post-facto maintenance
- Improvement-related maintenance
- Maintenance prevention

Three of these are especially important: preventive maintenance, improvement-related maintenance, and maintenance prevention.

Preventive Maintenance

Preventive maintenance is aimed at the prevention of breakdowns and defects. Daily activities include equipment checks, precision measurements, partial or complete overhauls at specified intervals, oil changes, lubrication, and so on. In addition, workers record equipment deterioration so they know to replace or repair worn parts before problems arise.

Recent technological advances in tools for inspection and diagnosis have enabled even more accurate and reliable equipment maintenance. The term *predictive maintenance* is used to describe activities that employ these advanced technologies.

Improvement-Related Maintenance

Improvement-related maintenance activities focus on improving the equipment to reduce future breakdowns or defects. They also make equipment easier to maintain.

When you understand a machine's weak points, you can make improvements designed to eliminate them. This, in turn, may facilitate checking, oiling, parts replacement and change-over, or other daily activities carried out by equipment operators.

Maintenance Prevention

In the development of new equipment, maintenance prevention is needed at the design stage. These activities aim at making equipment reliable, easy to care for, and user-friendly so operators can easily retool, adjust, and otherwise run it.

TOTAL PRODUCTIVE MAINTENANCE

Equipment management evolved from preventive to productive maintenance, but it was still an activity primarily practiced by the maintenance department. As such, it was never very successful at achieving zero breakdowns or zero defects. That is where TPM comes in. Based on small-group activities, TPM takes productive maintenance companywide, with the support and cooperation of managers and employees at all levels.

After all, the people most likely to first notice equipment abnormalities or other strange symptoms are not the maintenance staff but the operators who work with the equipment day in and day out. So the best way to prevent breakdowns is to have the operators give prompt notice of abnormalities and then have the maintenance staff promptly respond with corrective measures. Obviously, this cannot be done without active cooperation between both groups.

TPM, therefore, is characterized by production department workers participating in maintenance activities, an approach known as *autonomous maintenance.*

A full definition of TPM contains the following five points (see Figure 2):

1. It aims at getting the most efficient use of equipment (i.e., overall efficiency).
2. It establishes a total (companywide) PM system encompassing maintenance prevention, preventive maintenance, and improvement-related maintenance.
3. It requires the cooperation of equipment designers and engineers, equipment operators, and maintenance workers.
4. It involves every employee in the company.
5. It promotes and implements PM through small-group or team activities.

Figure 2. Definition of TPM

Bringing the whole company together behind TPM enables it to actually achieve goals such as zero breakdowns and zero defects, and these pay off in higher productivity and enhanced profitability.

THE FIVE PILLARS OF TPM

The following is a brief description of what are called the five pillars of TPM:

1. Improvement activities designed to increase equipment effectiveness. This is accomplished mainly by eliminating the six big equipment-related losses. (These will be explained later.)

2. An autonomous maintenance program to be performed by equipment operators. This is established as operators are trained to know their equipment.

3. A planned maintenance system. This increases the efficiency of traditional preventive maintenance activities.

4. Training to improve operation and maintenance skills. This raises the skill levels of equipment operators and maintenance workers.

5. A system for MP design and early equipment management. MP design helps create equipment that requires less maintenance. Early equipment management gets new equipment operating normally in less time.

While all five pillars are essential to TPM development, the first three are the most important for equipment operators. These will be described later in this book.

Figure 3 presents an overview of TPM. The elements will be explained in more detail in subsequent chapters, but for now this figure will provide an overall picture of TPM. It might also prove useful as a review summary to use after reading this book.

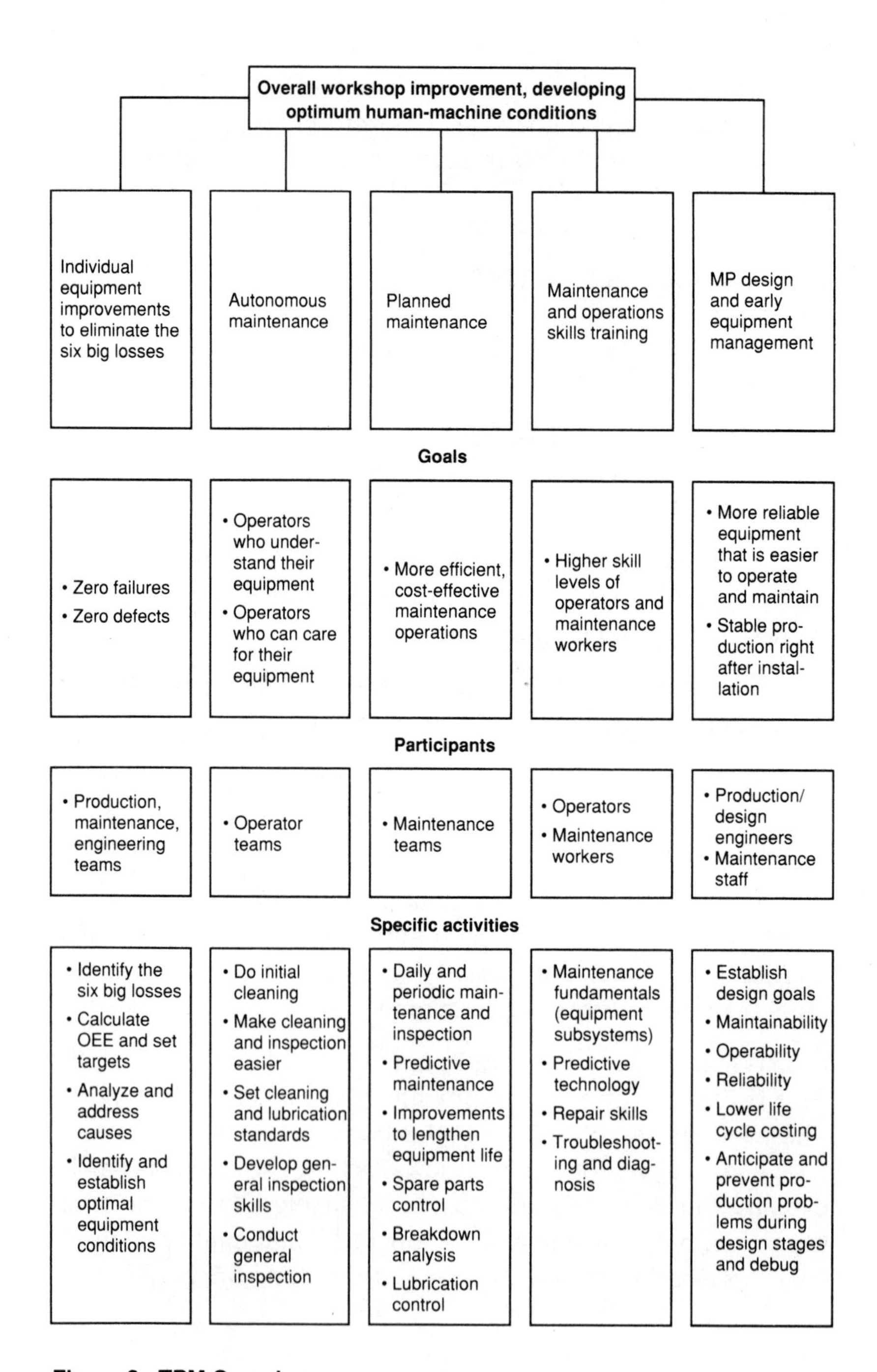

Figure 3. TPM Overview

3

Characteristics and Goals of TPM

THREE PRINCIPLES OF PREVENTION

One of the main characteristics of TPM is its aggressive pursuit of absolute goals, such as zero breakdowns and zero defects. In order to have "zero" anything, it must be prevented from happening even once. Therefore TPM emphasizes preventive action above all else. It is simply too late if you wait until a problem happens to fix it. In TPM, prevention is based on the following three principles:

1. ***Maintenance of normal conditions.*** To maintain normal operating conditions, operators prevent deterioration by cleaning, checking, oiling, tightening, and precision-checking the equipment on a daily basis.

2. ***Early discovery of abnormalities.*** While performing these activities, operators use their senses and measurement tools to detect abnormalities as soon as they appear. Maintenance workers also conduct periodic diagnostic tests to check for abnormalities using specialized tools.

3. ***Prompt response.*** Operators and maintenance staff respond immediately to abnormalities.

TWO MAIN GOALS OF TPM

The two main goals of TPM are to develop optimal conditions for the workshop as a human-machine system and to improve the overall quality of the workplace environment.

The Workshop as an Optimal Human-Machine System

Every workshop consists of a particular combination of two components: people (operators) and machines (equipment). No matter how these are combined — one operator per machine, one operator handling several machines, or a control board for automated machines, transfer machines, or robots —

the workshop system consists of people working closely with machines. These are called *human-machine* systems.

How well each human-machine system works to manufacture products depends mainly on how well human work meshes with the work of machines. Human work includes operating the machines correctly, adjusting, cleaning, oiling, replacing parts, etc., while machine work can include things like maintaining standard motions, precision, tolerances, etc.

The role of equipment operators also involves measuring how well the equipment is performing its functions and applying the results to improve its performance. While equipment actually makes the products, operators and maintenance staff maintain its health. To get maximum use of equipment, the ideal condition for each component must be known, as well as the measurement values that indicate peak performance. Once we know these conditions and values, our job is to maintain them. The more complicated the equipment is, the greater our responsibility becomes.

Unfortunately, few workshops come even remotely close to understanding and maintaining ideal conditions. Most have an awkward combination of people and machines, and a walk around practically any workshop will show equipment that is no longer in top shape because people have let it deteriorate. They do not maintain standards or carry out periodic measurements, so frequent equipment problems require disruptive and time-consuming repairs. Since most operators are not skilled in detecting equipment abnormalities, they seldom notice when equipment is giving signs of trouble. They take for granted that they must deal with problems only as they arise.

This state exists in most workplaces because operators are only responsible for operating the equipment and checking the quality of the products. They are not expected to be able to detect abnormalities. Similarly, maintenance workers view their role as simply repairing sporadic breakdowns. They are not concerned about defect loss, reduced speed, or other losses incurred by equipment operating in less-than-ideal condition. As long as it runs fairly well, they are satisfied.

Never forget: the main role in every human-machine system is the human one.

Keeping Equipment in Top Shape

Developing optimal conditions for the workshop as a human-machine system means finding the best combination of human and machine conditions.

For the machine side, that means getting the equipment into top shape. To do this, you must methodically identify and solve each problem that arises, and apply the TPM improvement approach. This requires the active participation of operators, maintenance staff, and production engineers.

There are two main points to remember about establishing an optimal human-machine system:

- Restore equipment to optimal operating conditions. Maintenance, production, and engineering must cooperate to thoroughly research the ideal conditions for each group of parts or equipment unit.
- Keep the equipment running at optimal operating conditions. In this regard, the particular responsibilities of operators, maintenance workers, and other technical staff must be carefully defined, agreed to, and carried out.

Improving the Quality of the Workplace Environment

Most workplaces are awash in loss, mainly because of quality defects and breakdowns. No matter how hard workers struggle to turn out good products, they always end up dealing with lots of defects and rework. Why do these losses occur? Because no one has done anything to change the workplace conditions that cause them. In other words, nothing can prevent quality defects and rework until the workplace itself is changed.

To change the workplace, there must be a change in the way of thinking about it and everything in it. For example, when a piece of equipment starts wearing down or vibrating, many operators either do not notice it or do not feel the need to do anything about it. But this must change. You need to concern yourself with even minor defects in equipment conditions, because the overall quality of the workplace, good or bad, is determined by human attitudes and behavior. So the most important thing for you to cultivate is an ability to discover even the most minor abnormalities *before* they develop into major problems.

Changing Our Attitudes

In many workshops, breakdowns, quality defects, and rework are always "someone else's fault." This is a big problem.

Quality improvement will not go anywhere until every employee takes responsibility for the workplace and sees each breakdown or defect as an embarrassment. Whenever a breakdown or defect occurs, we should seriously consider how we let such a thing happen.

Taking responsibility also means taking action. When your equipment breaks down, you should meet with a maintenance worker to find out exactly what caused it or what could have prevented it. Then you can figure out a way to keep the same thing from happening again. This is something you can do on your own initiative.

When each individual takes personal responsibility, the workplace will gradually improve.

TPM strongly emphasizes quality-consciousness among operators and improvement of the overall quality of the workplace environment. Still, attitudes are not easily changed. Changing the equipment, the people, and the entire workplace has to proceed step by step.

Equipment must be changed so it is sparkling clean and will not develop abnormalities so easily, and the workplace environment must be changed so its overall quality is improved. People, on the other hand, must change *themselves* so they understand and treat their work and workshop in a new and better way.

Changing the Equipment

In TPM, changing the equipment means changing how we think about our work. For example:

- Cleaning becomes inspection.
- Inspection reveals abnormalities.
- Abnormalities can be restored or improved.
- Restoration and improvement produce positive effects.
- Positive effects lead to pride in the workplace.

Here is a closer look at these changes.

Cleaning becomes inspection for abnormalities. Cleaning equipment requires you to touch and move it, coming into closer contact with it, making it easier to tell when it is acting abnormally. For example, while cleaning a machine component, you may notice something loose inside. Then you may wonder how it got that way through normal wear or perhaps because of a buildup of dust or other contaminants. You may also wonder what might happen if the part is left as it is.

Thus cleaning becomes a means of discovering abnormalities and dealing with them at an early stage. When an initial cleaning is done, it is not unusual to find anywhere from 10 to 50 abnormalities in each machine!

All abnormalities can be restored or improved. The next step is to understand the abnormalities that were discovered and come up with ways of restoring or improving them. With the help of maintenance, you can easily repair or replace some defective components. Perhaps the wiring layout needs better organization or a loose part needs tightening. Study the problems and try solutions.

Often this is much easier said than done. But it is only by struggling with puzzling problems and finding difficult solutions that you can come to appreciate the issues with which maintenance workers must deal. This experience also helps you learn not to allow troublesome problems to happen again.

Consider an oil pan that fills up soon after it is cleaned. You know how much trouble it is to clean the oil pan, so you wonder what fills it up so quickly. A close look at the fluid reveals a combination of cutting oil, hydraulic oil, and lubricant. With help from maintenance, you locate these leakage points and try various ways to stop the leaks, until you finally succeed. Having worked so hard, you naturally want to prevent the problem from recurring.

Restoration and improvement produce positive effects. Things that are restored or improved often start deteriorating again, unless we intervene. But this is no reason to give up. It's time to start thinking of ways to maintain the improvement, perhaps by using a different approach. When you make a successful improvement and find ways to maintain that improvement, you can be doubly proud. As the saying goes, the bigger the battle, the greater the glory.

From the viewpoint of the operator, "changing the equipment" involves cultivating a sharp eye for abnormalities, taking the trouble to fix them, experiencing the pleasure of making successful improvements, and finding ways to maintain them. Even a machine that seems no better than a rickety old junk heap can be cleaned and improved until it runs like a Swiss watch.

There is no joy or glory in going halfway; it takes determination to change equipment the TPM way.

Changing the Workplace

Once you have developed a sharp eye for abnormalities, there is no need to stop with the equipment. You can then focus on everything in the workplace — tools, processing conditions, control systems, and so on. These peripheral aspects of the workplace are also awash with abnormalities, but prompt action will prevent them from developing into big problems. The first step in changing the workplace is taken when you begin to notice and list the many things that need improvement.

Making improvements continually brings you new problems. For example, suppose that the cycle time needs to be shortened. This itself raises new issues. Dealing with them requires figuring out what can be done by the workers, what must be handled by the technical staff, and how the solution should be carried out. Soon you get used to this continuous improvement process, and it becomes a natural part of improving the overall quality of the workplace.

In summary, changing the equipment leads to changing attitudes and behaviors, which in turn leads to improving the overall quality of the workplace. These three changes — in the equipment, the people, and the workplace — are what TPM is all about.

4

Eliminating Equipment Losses

THE SIX BIG LOSSES: STUMBLING BLOCKS ON THE ROAD TO HIGHER EQUIPMENT EFFECTIVENESS

One of the goals of TPM is to improve equipment effectiveness, and this chapter describes the TPM approach to achieving that goal.

Basically, there are two ways to improve equipment effectiveness: a positive way and a negative way. The positive way is by making the most of the functions and performance features of the equipment. The negative way is by eliminating the obstacles to efficiency — obstacles that, in TPM, are called the six big losses. They are

1. Breakdown losses
2. Setup and adjustment losses
3. Idling and minor stoppage losses
4. Speed losses
5. Quality defects and rework
6. Start-up/yield losses (reduced yield between machine start-up and stable production)

In a TPM program, eliminating these specific equipment losses is the focus of equipment improvement project teams made up of production, maintenance, and engineering personnel. These teams identify key pieces of equipment in every work areas, measure the types of loss experienced, and then study carefully all factors that might be contributing to the loss related to equipment conditions, the material, the methods used by operators, and so on.

MEASURING EQUIPMENT EFFECTIVENESS

If you were told that equipment effectiveness at your plant was more than 85 percent, you might reasonably assume that the equipment is being operated efficiently and effectively. But how was this figure calculated and on what data were the calculations based?

Often, what is referred to as the rate of equipment effectiveness is actually the operating rate or *availability*, the time when equipment is up and running. TPM is not limited to dealing with breakdowns that affect availability, however; rather it raises the level of total equipment effectiveness by improving all related factors:

- *Availability:* Improved by eliminating breakdowns, setup/adjustment losses, and other stoppage losses
- *Performance:* Improved by eliminating speed losses, minor stoppages, and idling
- *Quality* (rate of quality products): Improved by eliminating quality defects in process and during start-up

Overall effectiveness can be measured using the formula:

Overall equipment effectiveness = Availability × Performance rate × Quality rate

Availability	Performance rate	Quality rate
• Breakdown losses • Setup and adjustment losses • Others	• Idling and minor stoppage losses • Reduced speed losses	• Quality defect and rework losses • Start-up losses

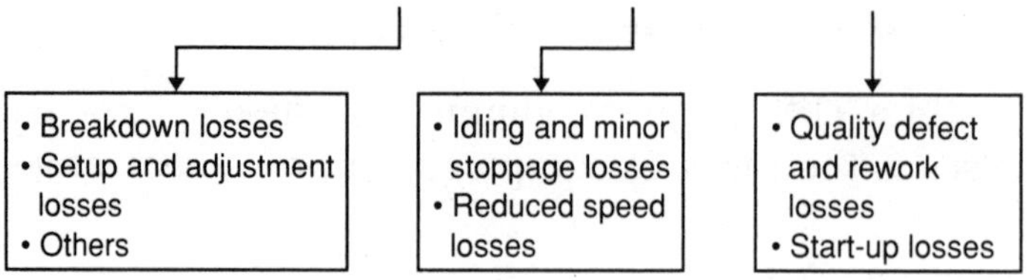

The availability, performance, and quality rates can be determined in each work center, but the importance of each factor varies according to the characteristics of the product, equipment, and production systems involved. For example, if adjustments and breakdowns are high, the operating rate will be low, and if many minor stoppages occur, the performance rate will be low. A high level of equipment effectiveness can be achieved only when all three rates are high.

AVAILABILITY (OPERATING RATE)

The operating rate tells us what percentage of time equipment is actually running *when we need it*. This is expressed in the following calculation:

$$\text{Availability (operating rate)} = \frac{\text{loading time} - \text{downtime}}{\text{loading time}} \times 100$$

In this case, loading time is the daily (or monthly) time available for operation minus all forms of scheduled stops — breaks in the production schedule, stoppages for routine maintenance, morning meetings, and so on. Downtime is the total time taken for unscheduled stoppages such as breakdowns, retooling, and adjustments. Loading time minus downtime yields the operating time.

For instance, suppose the loading time for a given day is 460 minutes, and downtime totals 100 minutes (60 minutes due to breakdowns, 20 minutes for retooling, and 20 minutes due to adjustments). Operating time for the day would therefore be 460 minus 100, or 360 minutes.

The availability (operating rate) can then be calculated as:

$$\text{Availability} = \frac{360}{460} \times 100 = 78\%$$

However, 78 percent availability does not accurately indicate actual operating conditions. It does not account for defects, reduced speed, and other loss factors.

PERFORMANCE RATE

The performance rate is based on the operating speed rate and the net operating time. The operating speed rate tells us how fast a machine is running (in terms of cycle time or strokes) compared to its ideal or designed speed. When the performance rate shows a speed reduction, it reflects a hidden loss. We use the following equation to calculate the operating speed rate:

$$\text{Operating speed rate} = \frac{\text{ideal cycle time}}{\text{actual cycle time}} \times 100$$

For example, if the theoretical (or standard) cycle time per item is 0.5 minutes and the actual cycle time per item is 0.8 minutes, the calculations would go as follows:

$$\text{Operating speed rate} = \frac{0.5 \text{ minutes}}{0.8 \text{ minutes}} \times 100 = 62.5\%$$

Net operating time is the time during which equipment is being operated at a constant speed within a specified period. Here the issue is not how fast the equipment is operating relative to the ideal speed, but whether it is running at a constant speed without interruption. Net operating time can therefore be used to calculate loss due to idling and minor stoppages, or other problems not usually mentioned in the daily log.

The formula for net operating time is as follows:

$$\text{Net operating rate} = \frac{\text{output} \times \text{actual cycle time}}{\text{loading time} - \text{downtime}} \times 100$$

For example, if the number of processed items per day is 400, the actual cycle time per item is 0.8 minutes, and the operation time is 400 minutes:

$$\text{Net operating rate} = \frac{\text{400 items} \times \text{0.8 minutes}}{\text{400 minutes}} \times 100 = 80\%$$

The remaining 20 percent represents losses caused by minor stoppages.

Now we will calculate the performance rate:

$$\text{Performance rate} = \text{net operating rate} \times \text{operating speed rate}$$

$$= \frac{\text{processed amount} \times \text{actual cycle time}}{\text{operation time}} \times \frac{\text{ideal cycle time}}{\text{actual cycle time}}$$

$$= \frac{\text{processed amount} \times \text{ideal cycle time}}{\text{operation time}}$$

$$= \frac{\text{400 (items)} \times \text{0.5 minutes}}{\text{400 minutes}} \times 100 = 50\%$$

or $(0.625 \times 0.80 \times 100 = 50\%)$

If the rate of quality products is 98 percent, then the overall equipment effectiveness is as follows:

$$\text{Availability} \times \text{Performance rate} \times \text{Quality rate}$$

or $(0.87 \times 0.50 \times 0.98 \times 100 = 42.6\%)$

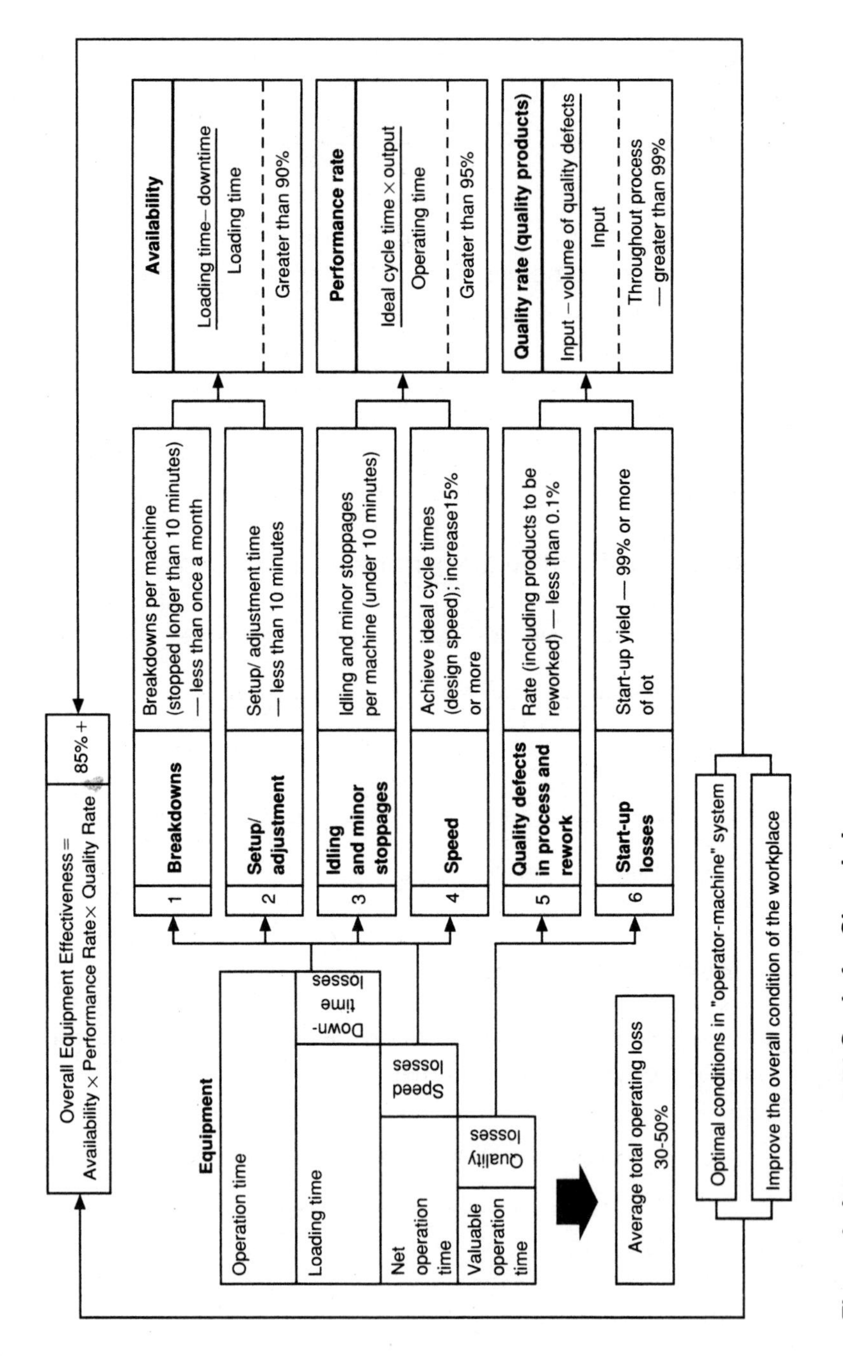

Figure 4. Improvement Goals for Chronic Losses

Even though the availability is 87 percent, the overall equipment effectiveness, when actually calculated, is not even 50 percent, but an astonishingly low 42.6 percent. This very poor overall effectiveness rating is due to poor operating speed and poor net operating time. So these calculations suggest the company should find ways to speed up the equipment and eliminate idling and minor stoppages.

TPM AND OVERALL EQUIPMENT EFFECTIVENESS

As you can see, there are many ways to calculate equipment loss. In TPM, the availability of equipment, its performance, and the product quality figures are all multiplied together to arrive at a measurement of the overall operating condition of the equipment. This can be used for all types of equipment.This method of calculation can also show a company which of the six big losses they need to concentrate on to increase equipment efficiency (see Figure 4).

Although the six big losses can be found in every workplace, the relative proportion of each will vary depending on equipment characteristics, line configuration, automation conditions, and other factors. For example, if the workplace has an abundance of setup/adjustment and breakdown loss, it will have an especially poor availability (operating rate). Likewise, a workshop loaded with idling and minor stoppages will have a particularly low performance rate.

Therefore, at any workplace, equipment improvement project teams first find out which losses have the greatest impact on equipment effectiveness, and then address the bulk of the improvement efforts toward those. To do this, teams follow these steps:

1. Measure the extent of each of the six big losses.
2. Determine how much each loss affects overall equipment effectiveness.

3. Find out what problems stand in the way of improving availability, performance rate, and quality rate.
4. Determine targets and orientations needed to solve problems discovered in step 3.
5. Find out how higher equipment effectiveness will affect cost-cutting and profit-boosting.

Breakdown Losses

Breakdowns are by far the biggest of the six big losses. There are two kinds: function-loss and function-reduction.

Function-loss breakdowns tend to occur sporadically (suddenly). They are easy to notice because they are relatively dramatic, for example, when a tool breaks or a motor burns out. On the other hand, function-reduction breakdowns enable the equipment to continue operating, but at a reduced level of efficiency; a simple example is a fluorescent lamp that begins to dim or flicker. Very often, these function-reducing failures can be discovered only by keen observation. But when they are overlooked, they give rise to idling and minor stoppages, rework, reduced speed, and other problems, and can become the cause of function-loss breakdowns.

Breakdowns are caused by all sorts of factors, but we usually notice only the big problems and overlook the many slight defects that also contribute to them. Obviously, the big problems deserve attention, but the slight defects deserve equal time because of their cumulative effect. Many breakdowns happen simply because seemingly minor things such as loose screws, abrasion, debris, and contaminants are ignored until, together, they affect the efficiency of the equipment.

The following seven actions will help you reach the target of zero breakdowns:

1. Prevent Accelerated Deterioration

Accelerated deterioration is simply deterioration that is artificially boosted by neglect, such as when equipment overheats because it is not oiled as often as it should be or when equipment is not checked and tightened. Looseness in one part soon affects others, producing a chain reaction eventually leading to a breakdown. Most breakdowns, in fact, are caused by accelerated deterioration. The first decisive step toward reducing breakdowns is to eliminate accelerated deterioration.

2. Maintain Basic Equipment Conditions

Three basic activities — cleaning, lubricating, and bolt tightening — must be performed to maintain basic equipment conditions and thus control deterioration. If these are not maintained, the workshop will certainly experience lots of breakdowns.

There are various reasons why workers fail to maintain basic equipment conditions. Sometimes they do not know how. Sometimes they do know how, but are not expected to do it. Those who do not know need to be taught — but they need to learn not only how to do the basic maintenance activities, but also why they are so important. Sometimes workers want to maintain basic equipment conditions but for some reason find it too difficult. For example, checking a machine may require complicated or time-consuming disassembly or a dangerous procedure such as work on a high ladder. In such cases, there is no choice but to improve the equipment so it can be maintained more easily.

3. Maintain Operating Conditions

Many breakdowns are caused by equipment that must "strain" to operate beyond its normal range because correct conditions are not maintained. Operating equipment under conditions exceeding the limits specified in the operating manual is practically asking for breakdowns. This is why maintaining correct operating conditions is so important.

4. Improve Maintenance Quality

Sometimes breakdowns occur in recently replaced or repaired parts because the maintenance worker did not possess the skills to perform the repair or installation correctly. To prevent these mistakes and improve maintenance quality, the skill levels of maintenance workers must be raised through training.

5. Take Repair Work Beyond Quick-Fix Measures

Repair work is usually focused only on getting the equipment up and running quickly, without much concern for the root causes of the breakdown. For example, if the most obvious cause was a broken bolt that held a cylinder in place, the repair work often consists of simply replacing the bolt without looking into why it broke. Obviously, such an approach only invites a recurrence of the same problem. What's missing is an attitude that seeks the root cause — which, admittedly, cannot always be found. Without this, however, the thorough maintenance required by TPM cannot exist.

6. Correct Design Weaknesses

One reason breakdowns become chronic is that there is not enough investigation into the weaknesses built into the equipment design. All too often design flaws are either not studied at all or not studied deeply enough to uncover their full implications. As a result, maintenance is not improvement-oriented, and breakdowns become chronic.

7. Learn as Much as Possible from Each Breakdown

Once a breakdown has occurred, be certain to learn everything you can. By studying the causes, conditions, and previous inspection and repair methods, much can be learned about how to prevent it from happening again, not only in the equipment at hand, but also in similar models.

It is unfortunate that the lessons to be learned from a breakdown are seldom put to good use. Quite often breakdown reports are filed away and forgotten instead of used for future reference. Learn to get the most out of such reference materials, because they can show maintenance workers and operators how to prevent future breakdowns.

Setup and Adjustment Losses

Setup and adjustment losses are stoppage losses occurring during setup procedures such as retooling, etc. Setup and adjustment time is the time required for stopping current production and setting up for production of the next part or product. Adjustments tend to use the greatest amount of this time.

Sometimes adjustments are required because of lack of rigidity or some mechanical deficiency. However, in trying to reduce their number, first look into the adjustment mechanisms, and divide adjustments into the avoidable (improvable) and the unavoidable (not improvable). At a typical factory 70 to 80 percent of adjustments are avoidable; these might include:

- Adjustments needed because of an accumulation of slight errors in precision, for example, repeated imprecise jig or equipment settings.
- Adjustments needed because measurement methods have not been standardized, or when standards are inconsistent.

Here are two steps that can be taken to eliminate the need for adjustments.

1. Review Precision Settings for Equipment, Jigs, and Tools

In many cases adjustments can be scaled down considerably simply by improving the precision settings of the equipment, jigs, and tools. This is because the accumulation of imprecise settings creates the need for many otherwise avoidable adjustments.

Whether the imprecision exists only in certain parts or is more widespread, you cannot eliminate it until you find out exactly where it is and how it can be corrected. Also, since the range of precision varies with each piece of equipment, you must carry out separate precision studies for each.

2. Promote Standardization

Lack of consistency in the standards for measurement, quantification, and other operation and maintenance procedures is another cause of unnecessary adjustments. The solution is to set clear, consistent, and precise standards for all procedures. In addition, promote standard tool usage as well as assembly and installation methods.

Idling and Minor Stoppage Losses

Unlike ordinary breakdowns, idling and minor stoppages are caused by temporary problems in the equipment. For example, a workpiece may jam in a chute, or a quality control sensor may temporarily shut down the equipment. As soon as someone removes the jammed workpiece or resets the sensor, it operates normally again. Therefore idling and minor stoppages are different from ordinary breakdowns, but they often interfere with efficiency, especially in automated equipment.

Because idling and minor stoppages can usually be restored quite simply, they tend to be overlooked and not regarded as loss. But they are, indeed, losses; this must be made obvious to everyone concerned. Until it is clearly understood just how much of a hindrance they really are, thorough measures to eliminate them cannot be devised.

In factories with many equipment units, each instance of idling or minor stoppage will require time to correct, but obviously the longer it takes, the greater the problem. Today there are more and more completely automated factories in which idling and minor stoppages pose a very big problem because no one is there to respond right away.

Remember the following three points when attempting to eliminate idling and minor stoppages:

1. Carefully Observe What Is Happening

Many attempts to eliminate idling and minor stoppages are stymied by knowing only the *results* of these events and not the phenomena *at the time* of the event. Very seldom does a machine idle or have a minor stoppage while someone is watching. Therefore you usually have to guess the conditions that were present on the basis of the results and then create a corrective measure based on that guess. If possible, it is much better to watch the equipment until it has another problem and only then plan a corrective measure.

2. Correct Slight Defects

Often slight defects are not even recognized as true defects and, even when noticed, are ignored. For example, a chute may clog because it is not properly swept out, or has a small dent in the side, or some other minor defect.

When dealing with idling and minor stoppages, it is especially important to look for these minor defects. Eliminating them can cut the number of idling and minor stoppages in half.

3. Understand the Optimal Conditions

Problems arise because people simply accept the current settings for their equipment without checking to see if they are optimal. Maintaining less than ideal conditions also invites idling and minor stoppages. Teams should take the time to review the settings and see if better ones can be established.

Speed Loss

Speed loss occurs when there is a difference between the speed at which a machine is designed to operate and its actual operating speed. Speed losses are typically overlooked in equipment operation, although they constitute a large obstacle to equipment effectiveness and should be studied carefully. The goal must be to eliminate the gap between design speed and actual speed.

Equipment may be run at less than ideal or design speed for a variety of reasons: mechanical problems and defective quality, a history of past problems, or fear of abusing or overtaxing the equipment. Often, the optimal speed is simply not known. On the other hand, deliberately increasing the operating speed actually contributes to problem solving by revealing latent defects in equipment conditions.

Quality Defects and Rework

Quality defects and rework are losses in quality caused by malfunctioning production equipment. In general, sporadic defects are easily and promptly corrected by returning equipment conditions to normal. These defects include sudden increases in the quantity of defects, or other dramatic phenomena. The causes of chronic defects, on the other hand, are difficult to identify. Quick fixes to restore the status quo rarely solve the problem, and the actual conditions causing the defects may be ignored or neglected. Defects that can be corrected through rework should also be counted as chronic losses and not ignored.

Eliminating chronic defects, like reducing chronic breakdowns, requires thorough investigation and innovative remedial action. The conditions surrounding and causing the defect must be determined and then effectively controlled. Complete elimination of defects is, as always, the main goal.

Since there are different types of defects — sporadic and chronic — reaching the goal of zero defects is all the more difficult. Reaching it requires coming up with measures based on a comprehensive understanding of all defects. There are four key points for eliminating quality defects:

1. Do not jump to conclusions about causes, and be sure corrective measures treat all causes being considered.
2. Carefully observe current conditions.
3. Review the list of causal factors.
4. Look again for slight defects, which are usually hidden among other causal factors.

Start-up/Yield Losses

Start-up/yield losses are those incurred because of the reduced yield between the time the machine is started up and when stable production is finally achieved. Often, start-up/yield losses are difficult to identify. Their extent depends on the stability of processing conditions, worker training, loss incurred by test operations, and other factors. In any case, this usually adds up to a lot of loss.

5

Autonomous Maintenance Activities in Production

The principal way in which the production department participates in TPM is through autonomous maintenance — cleaning, inspection, and simple adjustments performed by operators systematically trained through a step-by-step program.

AUTONOMOUS MAINTENANCE DEFINED

The purpose of an autonomous maintenance program is threefold. First, it brings production and maintenance people together to accomplish a common goal — to stabilize equipment conditions and halt accelerated deterioration. Operators learn to carry out important daily tasks that maintenance personnel rarely have time for. These tasks include cleaning and inspection, lubrication, precision checks, and other light maintenance tasks, including simple replacements and repairs in some environments.

Second, an autonomous maintenance program is designed to help operators learn more about how their equipment functions, what common problems can occur and why, and how those problems can be prevented by the early detection and treatment of abnormal conditions. Third, the program prepares operators to be active partners with maintenance and engineering personnel in improving the overall performance and reliability of equipment.

Traditionally, the general attitude on the shop floor has been, "I run it, you fix it." Operators were responsible only for setting up workpieces, operating the equipment, and checking the quality of processed work. All management of the equipment's condition was the responsibility of maintenance staff. By now it should be clear that this way of thinking does not promote optimal equipment performance.

The alternatives are sad indeed, because as operators you can easily prevent many breakdowns and quality problems by learning how to recognize abnormal conditions. A great deal of this learning can come about simply through your physical contact with the equipment — by taking a little time to tighten loose bolts, lubricating dry parts and cleaning away dirt, and by noticing dirt or grime on friction surfaces and switches — conditions that can shorten equipment life.

While these tasks are easy enough to do, in very few factories are they done well. Often you can find clogged drains, empty oil supply equipment, and other results of neglect.

Autonomous maintenance teaches you, the equipment operator, to understand your equipment. Equipment knowledge is no longer limited to operation; now it also includes a lot of things traditionally regarded as maintenance work. This approach is becoming increasingly important as factories introduce more robots and automated systems. Most important, you need the ability to look at the quality of the products and the performance of the equipment and notice when something is not right.

This depends on the following three skills:

1. Knowing how to distinguish between normal and abnormal conditions (the ability to establish equipment conditions).

2. Knowing how to ensure that normal equipment conditions are met (the ability to maintain equipment conditions).
3. Knowing how to respond quickly to abnormalities (the ability to restore equipment conditions).

When you have mastered all three skills, you will understand the equipment well enough to recognize the causes of future problems. You will realize when the machine is about to produce defects or break down. You will also be able to respond quickly. The following list describes some of the skills operators need.

1. The ability to detect, correct, and prevent equipment abnormalities and make improvements. This includes understanding the importance of

- proper lubrication, including correct lubrication methods and methods for checking lubrication.

- cleaning (inspection) and proper cleaning methods.
- improving equipment to reduce the amount of debris and prevent its accumulation and spread.
- improving operation and maintenance procedures to prevent abnormalities and facilitate their prompt detection.

2. ***The ability to understand equipment functions and mechanisms, and the ability to detect causes of abnormalities.***

- Knowing what to look for when checking mechanisms.
- Applying the proper criteria for judging abnormalities.
- Understanding the relations between specific causes and abnormalities.
- Knowing with confidence when the equipment needs to be shut off.
- Being able to diagnose the causes of some types of failures.

3. ***The ability to understand the relationship between equipment and quality, and the ability to predict problems in quality and detect their causes.***

- Knowing how to conduct a physical analysis of a problem.
- Understanding the relationship between product quality characteristics and equipment mechanisms and functions.
- Understanding tolerance ranges for static and dynamic precision, and how to measure such precision.
- Understanding the causes of quality defects.

4. ***The ability to make repairs.***

- Ability to replace parts.
- Understanding of life expectancy of parts.
- Ability to deduce causes of breakdowns.
- Ability to take emergency measures.
- Ability to assist in overhaul repairs.

Obviously, anyone who masters all these skills has achieved a very high level indeed, and no one is expected to do that quickly. Instead, each skill should be studied and practiced for whatever time it takes to acquire proficiency.

IMPLEMENTING AUTONOMOUS MAINTENANCE IN SEVEN STEPS

Table 5-1 outlines the seven developmental stages of an autonomous maintenance program. These stages or steps are based on the experiences of many companies that have successfully implemented TPM. They represent an optimal division of responsibilities between production and maintenance departments in carrying out maintenance and improvement activities.

A Step-by-Step Approach

It is very difficult to do several things at the same time. That's why autonomous maintenance training takes a step-by-step approach, making sure each key skill is thoroughly learned before going on to the next. Autonomous maintenance is implemented in seven steps:

- Step 1. Conduct initial cleaning and inspection
- Step 2. Eliminate sources of contamination and inaccessible areas
- Step 3. Develop and test provisional cleaning, inspection, and lubrication standards
- Step 4. Conduct general inspection training and develop inspection procedures
- Step 5. Conduct general inspections autonomously
- Step 6. Organize and manage the workplace
- Step 7. Ongoing autonomous maintenance and advanced improvement activities

Table 1. Seven Steps of Autonomous Maintenance Activities

Step	Activity	Goals for Equipment (workplace diagnosis)	Goals for Group Members (TPM group diagnosis)
1. Conduct initial cleaning	Thoroughly remove debris and contaminants from equipment (remove unused equipment parts)	• Eliminate environmental causes of deterioration such as dust and dirt; prevent accelerated deterioration • Eliminate dust and dirt; improve quality of inspection and repairs and reduce time required • Discover and treat hidden defects	• Develop curiosity, interest, pride, and care for equipment through frequent contact • Develop leadership skills through small group activities
2. Eliminate sources of contami-nation and inaccessible areas	Eliminate the sources of dirt and debris; improve accessibility of areas that are hard to clean and lubricate; reduce time required for lubrication and cleaning	• Increase inherent reliability of equip-ment by preventing dust and other contaminants from adhering and accumulating • Enhance maintainability by improving cleaning and lubricating	• Learn equipment improvement concepts and tech-niques, while implementing small-scale improvements • Learn to participate in improvement through small group activity • Experience the satisfaction of successful improvements
3. Develop cleaning and lubrication standards	Set clear cleaning, lubrication, and inspection standards that can be easily maintained over short intervals; the time allowed for daily/periodic work must be clearly specified	• Maintain basic equipment conditions (deterioration-preventing activities): cleaning, lubrication, and inspection	• Understand the meaning and importance of mainte-nance by setting and maintaining our own standards (What is equipment control?) • Become better team members by taking on more responsibility individually
4. Conduct general inspection skills training	Conduct training on inspection skills in accordance with inspection manuals; find and correct minor defects through general inspections; modify equipment to facilitate inspection	• Visually inspect major parts of the equipment; restore deterioration; enhance reliability • Facilitate inspection through innovative methods, such as serial number plates, colored instruction labels, thermotape gauges, and indicators, see-thru covers, etc.	• Learn equipment mechanisms, functions, and inspec-tion criteria through inspection training; master inspection skills • Learn to perform simple repairs • Leaders enhance leadership skills through teaching; group members learn through participation • Sort out and study general inspection data; understand the importance of analyzed data

5. Conduct inspection autono-mously	Develop and use autonomous main-tenance check sheet (standardize cleaning, lubrication, and inspection standards for ease of application)	• Maintain optimal equipment conditions once deterioration is restored through general inspection • Use innovative visual control systems to make cleaning/lubrication/inspection more effective • Review equipment and human factors; clarify abnormal conditions • Implement improvements to make operation easier	• Draw up individual daily and periodic check sheets based on general inspection manual and equipment data and develop autonomous management skills • Learn importance of basic data-recording • Learn proper operating methods, signs of abnormality, and appropriate corrective actions
6. Organize and manage the workplace	Standardize various workplace regu-lations; improve work effectiveness, product quality, and the safety of the environment: • Reduce setup and adjustment time; eliminate work-in-process • Material handling standards on the shop floor • Collecting and recording data; standardization • Control standards and procedures for raw materials, work-in-process, products, spare parts, dies, jigs, and tools	• Review and improve plant layout, etc. • Standardize control of work-in-process, defective products, dies, jigs, tools, measuring instruments, material handling equipment, aisles, etc. • Implement visual control systems throughout the workplace	• Broaden the scope of autonomous maintenance by standardizing various management and control items • Be conscious of the need to improve standards and procedures continuously, based on a standardization practice and actual data analysis • Managers and supervisors are primarily responsible for continuously improving standards and procedures and promoting them on the shop floor
7. Carry out ongoing autonomous maintenance and advanced improvement activities	Develop company goals; engage in continuous improvement activities; improve equipment based on careful recording and regular analysis of MTBF	• Collect and analyze various types of data; improve equipment to increase reliability, maintainability, and ease of operation • Pinpoint weaknesses in equipment based on analysis of data, implement improvement plans to lengthen equip-ment life span and inspection cycles	• Gain heightened awareness of company goals and costs (especially maintanance costs) • Learn to perform simple repairs through training on repair techniques • Learn data collection and analysis and improvement techniques

These activities are carried out by operator teams with resource support and training provided by maintenance personnel, production managers, and engineering staff.

Each stage in the implementation of autonomous maintenance emphasized different developmental activities and goals, and each builds upon thorough understanding and practice of the previous step. Step 1 (initial cleaning and inspection), step 2 (eliminating sources of contamination and inaccessible areas), and step 3 (developing cleaning, inspection, and lubrication standards) promote the establishment of basic equipment conditions that are essential to effective autonomous maintenance.

Step 1: Initial Cleaning

In this first step, operator teams put the "cleaning is inspecting" motto into practice and confirm it with their own experience. The physical act of touching the equipment and shifting it around reveals abnormalities. You'll use all your senses to detect looseness and vibration, wear, misalignment, deflection, abnormal noise, overheating, and oil leaks. In any workshop, cleaning will reveal numerous slight abnormalities. Many will be the kind that by themselves or in combination can lead to major breakdowns or other losses unless detected early.

Sometimes cleaning also reveals big surprises, such as a cracked frame that had been masked by accumulated grime, a lubrication inlet also hidden by dirt, or switches so covered with grime they no longer function correctly.

Inadequate cleaning is often the cause of equipment and quality problems. Here are just a few examples:

- foreign particles in sliding parts, hydraulic or electrical systems produce frictional resistance, wear, clogging, leaks, and electrical faults that can lead to losses in precision, equipment malfunctions and failures.

- Dirt on chutes and workpieces in automated equipment can affect the flow of work, cause malfunctions and minor stoppages
- Contamination of plastic molding machine dies or the feed materials can cause carbonizing or resin leakage and burning, which affects quality and makes changeovers difficult
- In precision machining, dirt on jigs, tools, and mountings can cause eccentricity during machining and produce defective parts.
- In electroplating, contaminated workpieces or dirt in the electrolyte can cause plating defects.

Thorough cleaning means taking equipment apart to clean internal parts that you may never have seen, so this is a kind of inspection that naturally leads to discovering abnormalities. Over time you will learn the correct way to inspect equipment for abnormalities, distinguish between abnormalities and normal conditions, and look for causes. You should learn to look for things not ordinarily discovered when operating the equipment.

Find Sources of Contamination

If the equipment becomes dirty soon after cleaning, look for the source of contamination. Often a source of dirt or oil leakage will not be visible unless the equipment is thoroughly cleaned at least once a day, and this time-consuming task will itself prompt you to look for ways to reduce or eliminate the contamination. Ask yourself why the equipment should continue to leak and how you can prevent it.

If equipment is new and managed properly, initial cleaning will do little to expose hidden abnormalities. But initial cleaning is still valuable for understanding how various parts function. It will give you a better grasp of the whole series of motions and processes involved in the equipment operation.

Correct Problems Yourself Whenever Possible

When an operator finds an abnormality, he or she should tag its location. The team then needs to figure out which abnormalities they can correct themselves and which must be looked into by a maintenance technician, and they should set a deadline for correcting each one. However, teams should take care of as many as possible by themselves.

Step 2: Eliminate Contamination Sources and Inaccessible Areas

In this step operator teams make cleaning and inspecting easier by controlling the sources of dirt and grime and other forms of contamination they found during initial cleaning.

Control the Sources of Dirt

Teams begin with a goal of stopping contamination at its source, for example, by repairing a leaking pipe joint or reducing lubricant volume to prevent an overflow. Sometimes this is not possible, for example, when a blade continuously produces cutting debris or when a stream of coolant scatters debris and coolant away from the processing point. In such cases, teams can at least minimize the scattering by installing shields as close to the source as possible to localize it.

Make Cleaning and Inspection Easier

Sometimes the positioning of parts or the design of covers makes areas of equipment hard to clean and check. For example, if a filter-regulator-lubricator set is installed too close to the floor, it becomes hard to remove the filter to check the lubricator. Repositioning the equipment will facilitate inspection. Similarly, if V-belts require checking, make the cover easier to remove or, better, install a window in the cover.

Learn Through Trial and Error

During the improvement process you should expect to make some mistakes, so test your ideas through trial and error and simulation. For example, in developing localizing improvements, teams often install temporary shields made out of cardboard. Then they run the machine for a while to see how the debris scatters and how effectively the temporary shield contains it. They may have to test more than one temporary shield before recommending a design for the permanent fixture. Experimenting on the shop floor in this way not only builds your skills and confidence, it also avoids the cost of expensive engineering solutions that don't work out.

Keep in mind the following list of key points when making improvements:

- Make the equipment easier to clean.
- Minimize the spread of dirt, dust, and grime.

- Stop contamination at its source.
- Minimize the scattering of cutting oil and debris.
- Speed up the flow of cutting oil to prevent the accumulation of debris.
- Minimize the area through which cutting oil flows.
- Make the equipment easier to inspect.
- Install inspection windows in the equipment.
- Tighten loose areas in the equipment.
- Eliminate the need for oil pans.
- Install more oil gauges.
- Change the locations of the lubrication inlets.
- Change the lubrication method.
- Simplify wiring and piping layouts.
- Make it easier to replace equipment parts.

Step 3: Develop and Test Provisional Cleaning, Inspection, and Lubrication Standards

In this step, teams use experiences from the first two steps to determine the optimal cleaning and lubrication conditions for the equipment and to draft preliminary work standards to maintain those conditions. Work standards specify what needs to be done, where, why, how, and when. You must decide which parts of the equipment need cleaning and how often, which methods to use, what to inspect while cleaning, how to judge whether conditions are normal, and so on (see Figure 5). Having these standards will help you do your cleaning tasks with greater confidence and ability.

Having accepted responsibility for the equipment, teams should decide for themselves how to maintain it. Many workshops have already established cleaning and lubrication standards, but very few follow them thoroughly, mainly because the people who set the standards are seldom the ones who must apply them. Cleaning and lubrication standards are more likely to be followed when these three criteria are observed:

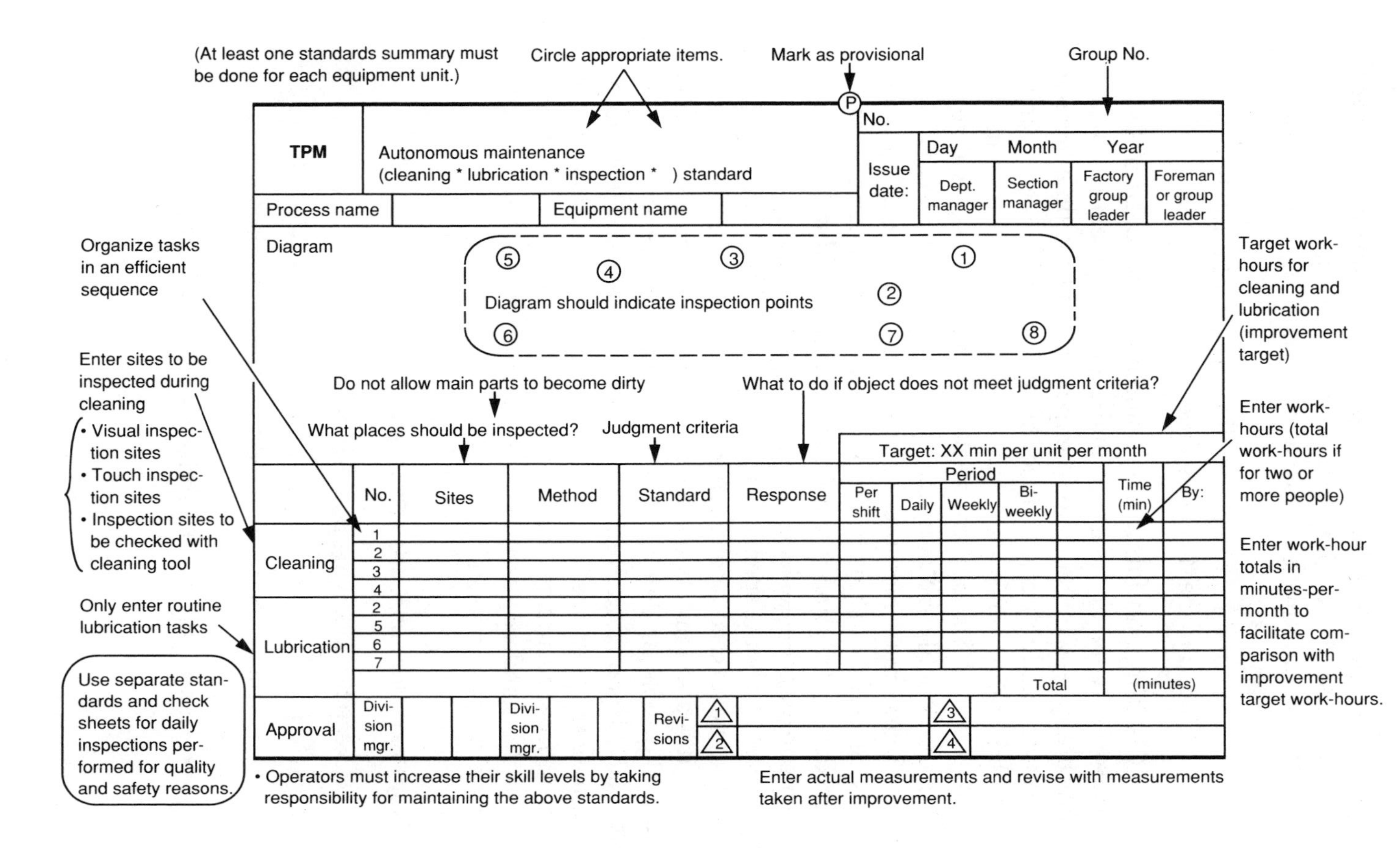

Figure 5. Cleaning and Lubrication Standards Summary for Autonomous Maintenance

1. The people *doing* the cleaning and lubricating understand the vital importance of these tasks.
2. The equipment is improved to make cleaning and lubrication easier.
3. The time required for cleaning and lubrication is included in the daily schedule.

Cleaning and lubrication standards will not be followed unless everyone in the group understands — theoretically and practically — why they are so important. But the time given to cleaning and lubrication cannot be unlimited. To keep cleaning and lubrication times short, you need to devise time-saving improvements. Without them, some maintenance tasks will not be done thoroughly and others will be omitted altogether.

Teams must measure the time required for each task and devise ways to shorten the more time-consuming ones. In scheduling daily and weekly maintenance, avoid imbalances by reorganizing the tasks or making time-saving improvements. Overall, the hours spent on cleaning, inspection, and lubrication should not exceed 2 percent of the total work-hours.

It is also important to incorporate the lessons learned from breakdowns or defects. Even when a breakdown was "inevitable," you can still find its causes, devise better ways to diagnose the abnormal signs that are its symptoms, and prevent them from recurring. This information must be incorporated into the standards as it becomes known, in the form of additional inspection items.

Key Points for Lubrication Standards

Keep the following points in mind:

- Clearly specify the lubricant to be used, and unify types whenever possible to reduce variety and promote consistency.
- Thoroughly list all lubrication inlets and other sites.

- In centrally lubricated equipment, improve the lubrication system and diagram the route from pump to main pipes, branch valves, branch pipes, and lubrication points.
- Check for blockage in branch valves and differences in branched volumes, and find out if lubricant is reaching all lubrication points.
- Measure the lubricant consumption rate (during one day or one week).
- Measure the amount used per application.
- Measure the pipe lengths (especially grease pipes) to see, for example, if two pipes may be needed instead of one.
- Review the disposal method for used lubricant.
- Clearly label all lubrication points.
- Establish a service station (for maintenance of lubricants and lubrication equipment).
- List all difficulties concerning lubrication and address them systematically.
- Work out the division of lubrication-related responsibilities with the maintenance department.

Step 4 (general inspection training) and step 5 (autonomous inspection) stress thorough equipment inspection and subsequent maintenance and standardization. Furthermore, these steps promote the development of strong observation and diagnostic skills in operators. During these phases, your company is likely to see substantial reductions in equipment failures.

Step 4: Conduct General Inspection Training

For you to understand your equipment you must learn about the common systems all types of equipment share, as well as the unique features of each equipment unit.

In this fourth stage, you receive basic instruction in equipment subsystems, such as lubrication, equipment elements (tightening nuts and bolts), pneumatics, hydraulics, electrical circuits, drive systems, and other basic technologies such as waterproofing and fire prevention, or others specific to the operation of your particular industry. You will use this knowledge while inspecting your equipment for abnormalities.

In pneumatics, for example, you need to understand the functions and structure of FRLs (filter, regulator, and lubricator sets) and how to adjust the volume of lubricant. This is valuable when cleaning and checking pneumatic equipment. Such training enables you to perform inspections knowing the particular points you must check and the most important maintenance management points.

General Inspection Skills Training

Typically, while operator teams are busy implementing the first three steps, maintenance and other technical staff prepare text materials, diagrams, cutaway models, and other training aids for the general inspection skills training that all operators receive. The following model is an efficient way of transferring this knowledge. Basic training is first provided to operator team leaders, who then pass on this knowledge one point at a time to team members, over an extended period. Team members apply new points as soon as they learn them by conducting focused inspections for new problems to address. They learn about one equipment subsystem at a time thoroughly (e.g., hydraulics), before tackling another topic (such as electrical systems). Teams also learn how to develop and use their own visual controls to make maintenance and inspection procedures easier and error-free.

Specifically, you will learn the structure of your equipment and its various systems, its functions, proper adjustments and use, structural problem points, and daily checkpoints. This is taught in step 4 because it builds on your earlier experience. If you were not already so familiar with your equipment, these lessons in basic maintenance would not be nearly as effective or useful to you. For example, up to this point, your methods for discovering abnormal conditions have been limited to using the five senses. With a broader understanding of equipment structure and function you can discover abnormalities using logical analysis as well as observation and common sense.

Consider the simple example of a bolt. Common sense tells you if the bolt is tight, but you must be trained to know that it is properly tightened only when the correct amount of torque has been used. To apply the correct torque rating to make best use of the bolt, you apply not only common sense but also theoretical knowledge.

To make sure you have learned more than just intellectual knowledge, you'll have the chance to discover its true value by practicing it every day. As you study and inspect each subsystem, you'll incorporate the checkpoints introduced during these lessons into your provisional standards manual.

Applying visual control. As the term *visual control* suggests, these are visual cues or indicators such as labels or color coding used to check equipment or detect abnormal conditions. They should be obvious enough for anyone to see and understand. Keep in mind the following points when developing visual controls:

- Is it easy to see what needs to be checked?
- Does everyone understand the function and structure of this part of the equipment?
- Is it easy to tell whether the condition being checked is normal or abnormal? (If not, have concrete criteria for judging been established?)
- Is it easy to know what action should be taken if the condition is abnormal?

One common practice for visual control is placing match marks on all critical bolts (see Figure 6). Having understood the function of each nut and bolt pair, you tighten the bolts to their specified torque and then paint a line across bolt, nut, and washer to the equipment body. In this way you can see at a glance whether a bolt has loosened. However, distinctions should be made among certain bolts. Bolts for which tightness is particularly critical must be prevented from loosening by using appropriate locking devices. Only after those measures are taken should they be given match marks.

Here are some ideas for visual control:

Lubrication

- Color-coded marks to indicate oil inlets
- Oil level and supply interval labels
- Indication of upper and lower oil level limits
- Indication of oil consumption per standard time unit
- Color-coding on oil cans to indicate oil types

Equipment elements

- "Inspected" marks and match marks
- Color-coding (blue) on bolts reserved for adjustment by maintenance personnel only
- Color-coding (yellow) on holes that do not require bolts

Pneumatics

- Pneumatic pressure gauge limits
- Oil level display
- Display of upper and lower oil level limits
- Labels on solenoids
- Tube connection marks (INLET, OUTLET)
- Color coded supply lines
- Flow direction (arrows)

Hydraulics

- Hydraulic pressure gauge limits
- Oil level display
- Oil type display

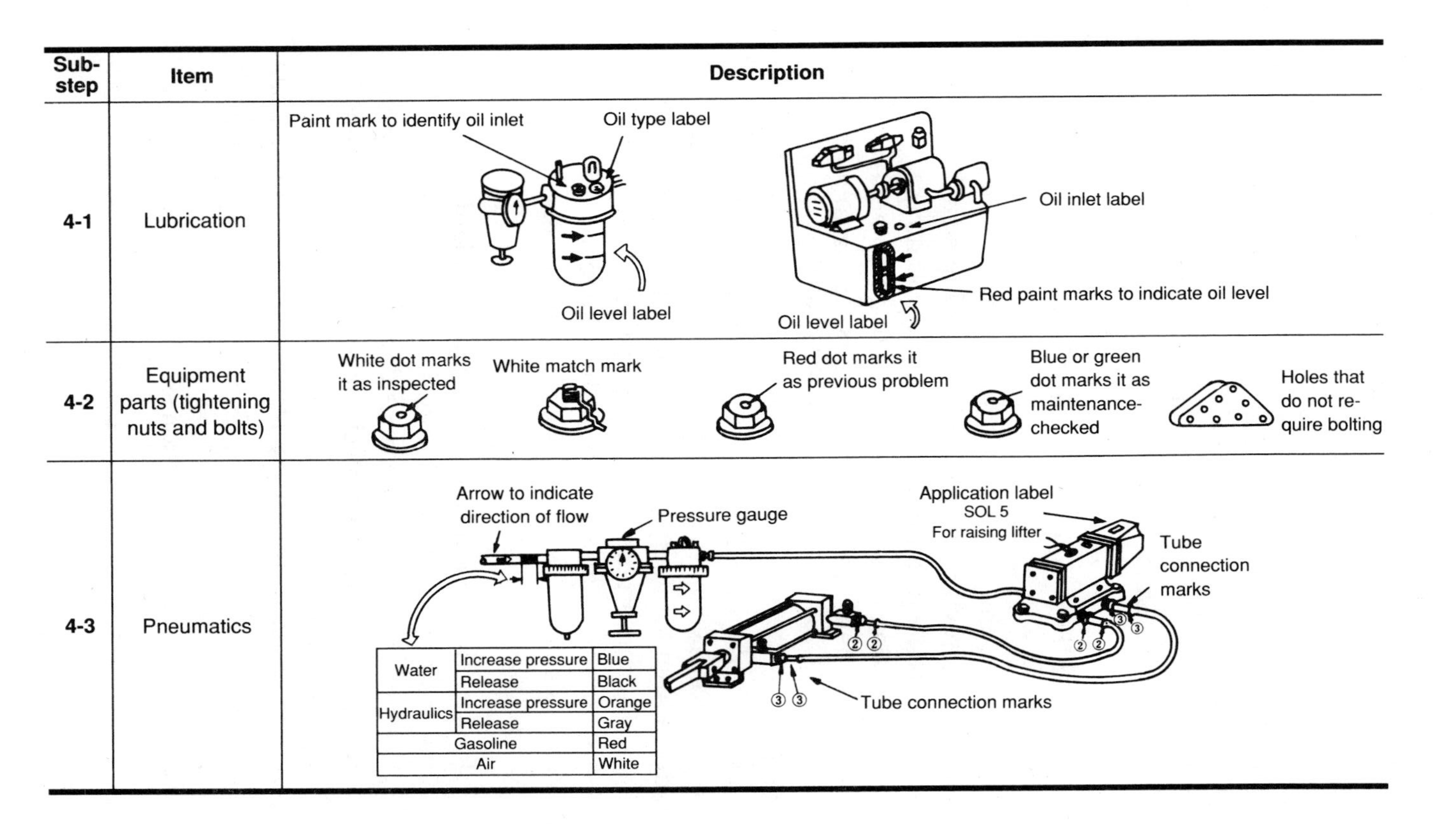

Figure 6. Examples of Visual Control

- Thermal label for hydraulic pump
- Solenoid labels
- Locknut match marks on relief valve

Drive systems
- Indicate V-belt/chain type
- Indicate V-belt/chain direction of rotation
- Install peep window for checking belts

Step 5: Conduct Inspection Autonomously

In this step, an overall inspection process is formalized by combining the provisional standards created in step 3 with the additional check items for routine general inspection.

All inspection items for each piece of equipment are split into two lists: items that can be handled using autonomous inspection and items requiring inspection by maintenance specialists.

If any breakdowns have occurred, work with the maintenance staff to develop inspection points that will prevent the breakdown from happening again and that can be performed by operators. Then incorporate these new inspection points into the standards. All inspection items must then be reviewed to make sure the work can still be done within reasonable time limits.

At this stage the activities are as follows:

- Review the item, method, and time standards for cleaning, inspection, and lubrication.
- Consult with the maintenance department about inspection points, and make specific and clear job assignments to avoid omissions.
- Check whether or not the inspection tasks can be done within scheduled work-hours, and make time-saving improvements if necessary.
- Check to see if the inspection skill level can be raised.

- Make sure autonomous inspection is carried out correctly by all operators.

Why do we go to this much trouble to develop autonomous inspection procedures? No matter how successfully equipment improvement teams establish the conditions for zero breakdowns and zero defects, unless you enforce a schedule of daily inspection, lubrication, and precision checks, breakdowns and defects will return. In other words, how well autonomous inspection is performed determines how permanent the improvements will be. That is why you cannot afford to neglect autonomous inspection or the need to develop a thorough understanding of your equipment.

Steps 6 and 7 (workplace organization and ongoing autonomous management) stress improvement activities informed by operators' increasing knowledge and experience and extending beyond the equipment to its surrounding environment. These activities increase involvement, skills, and cooperation among divisions. Operators become strongly identified with company goals and assume responsibility for the maintenance and improvement activities that are essential for effective self-management on the shop floor.

Step 6: Organize and Manage the Workplace

Once equipment conditions are under control, your shop-floor team activity can expand beyond the equipment to other aspects of the work environment. At this point teams often begin eliminating clutter and unnecessary items around their work areas and then organize what remains. Using simple principles of housekeeping and visual control, they establish standard quantities and locations for the essential elements: materials, work-in-process, and the process flow itself; tools, fixtures, and measuring instruments; operating standards — including setup and changeover procedures; and quality. These organizing and standardizing activities emphasize the following points:

- Decide when, by whom, and how each item should be used.
- Check the quality and quantity of items so that when needed they can be used to full advantage.
- Arrange items so that people can see at a glance where things are and how they should be used.
- Decide how to arrange tools and materials and determine what quantity is required according to their frequency of use.
- Store items so that they will occupy as little space as possible and can be moved easily.
- Decide who is responsible every day for management tasks and how materials, parts, or tools will be supplied or discarded.

These activities link equipment management and production management goals and promote transparency in the workplace. In other words, all standards become easy to understand and follow; all potential errors or abnormal conditions become easy to prevent or detect.

Of course, your team will also continue to refine standards for equipment inspection, cutting the time and making the checkpoints easier to find and to judge. You'll also work in cooperation with maintenance and engineering staff on the other 40 to 60 percent of problems that were not solved by restoration and improved maintenance procedures. You'll want to document the characteristics of continuing problems and add inspection items to control newly identified factors; you may also need to adjust standard values that turn out to be too loose.

Step 7: Carry out Ongoing Autonomous Management and Advanced Improvement Activities

Step 7 is less a step than a jumping off point for continuous equipment-focused improvement activity. Operator teams in co-operation with maintenance continue to refine the inspection process and to generate improvements that increase equipment life and effectiveness. You'll become increasingly involved with maintenance in gathering and analyzing equipment data such as the results of daily inspection, downtime statistics, oil and grease usage, quality defect data, tool wear records, and so on. And you will continue to build your analytical and diagnostic skills by working on increasingly challenging improvement projects that reflect evolving company goals, such as reliability and maintainability improvement or quality activities. For example, at this stage, as operators become full partners in the equipment management process and zero unplanned downtime starts to become a reality, all teams can focus on the goal of zero defects — achieving quality objectives through equipment-related tactics.

CHARACTERISTICS OF AUTONOMOUS MAINTENANCE PROGRAMS

Every factory must have a plan for autonomous maintenance activities that meets the specific goals of the enterprise and the unique needs of equipment and workshop personnel. Effective programs, however, have certain characteristics in common. This section reviews some of those features.

Organization-Led Activities

Although the word *autonomous* implies that small groups perform these activities entirely on their own, in fact, they receive considerable guidance and support from the companywide or plantwide TPM organization through a system of overlapping TPM planning and improvement teams involving everyone from the plant manager to operators and maintenance workers on the shop floor. In other words, shop-floor groups meet with their chosen team leaders; team leaders meet regularly with middle-management groups involving supervisors or area managers, and so on, depending on the nature of the plant organization. The higher-level groups provide direction, support, and feedback to the groups below them. This ensures that everything done within the framework of autonomous maintenance is consistent with other TPM plans and activities carried out plantwide and that necessary resources are authorized and available when needed to support those activities.

Use of Audits

To ensure consistent results and maximize group efforts, autonomous maintenance activities are audited at intervals to determine each group's readiness to go on to the next step. Your team should request an audit when it feels ready. In filling out the request form, you should describe what problems have taken most of the group's efforts and what improvements you emphasized (see Figure 7).

There are three important reasons for these audits:

1. To determine whether each step has been fully implemented.
2. To provide feedback to groups on the strengths and weaknesses of their activities.
3. To clarify what needs to be achieved and how best to achieve it.

Sometimes your group will pass on the first audit and sometimes only after the second or third. But the important

Audit Request Form

TPM Autonomous Maintenance Audit Request Form

Dept.

Procedural step to be audited: Step 1

To be completed by applicant

<table>
<tr><td>Workshop making request</td><td>Bearing factory, Dept. 4
Group 2, Section 1</td><td colspan="2">Equipment to be audited</td></tr>
<tr><td>Name of small group</td><td>"Our Gang"</td><td colspan="2">No. of equipment units to be audited</td></tr>
<tr><td>Name of leader
(No. of people in group)</td><td>Dan Graham, Men: 2
Women: 2 Total: 4</td><td rowspan="4">Activities to date</td><td>Item / No. of repeats</td></tr>
<tr><td>Self-evaluation score (date)</td><td>85 points (February 9)</td><td>1</td></tr>
<tr><td rowspan="2">Desired audit date</td><td rowspan="2">February 12, about 1:00pm</td><td>2</td></tr>
<tr><td>3</td></tr>
<tr><td>Result indicators</td><td colspan="3">No. of idling and minor stoppage incidents, rework rate, changes in changeover time</td></tr>
<tr><td>Points of emphasis for audit</td><td colspan="3">1. Group activities (No. of meetings prior to this audit)
Cleaning takes a lot of time because of the large number of machines in our section. We also spent a lot of time discussing ways to locate abnormalities and put a lot of emphasis on discovering them.
2. Workshop audit
We would like to find out if we overlooked any abnormalities. We would also like an evaluation of the improvement we made to localize the scattering of honing fluid.</td></tr>
</table>

To be completed by auditors

<table>
<tr><td>Audit date/time</td><td>February 12, 1-3pm</td><td>Audit team meeting place</td></tr>
<tr><td>Audit team members</td><td colspan="2">McKenzie, Jones & Rodriguez</td></tr>
</table>

Figure 7. Audit Request Form

thing is to change the way you think about your equipment and to sharpen your observation skills.

Use of Activity Boards

Although activity boards are a common tool for most small-group activities, they are rarely used effectively and quite often function simply as display boards. Their original purposes are as follows:

- To describe the strategy and orientation of your group; that is, your overall approach and methods.
- To describe the specific activities you've undertaken and to show how far they have progressed so it is easy to see what your group is trying to achieve, by what date, and in what manner.
- To post results of statistical trends for each of the six big losses; the overall equipment effectiveness rates: availability, performance, and quality; and other measures such as number of maintenance calls, lubricant consumption, cleaning times, etc. All these should also describe the relationship between your group's activities and its accomplishments.
- To describe key issues addressed as well as the reasons for immediate actions and an indication of the next key issue to be confronted.
- To record issues to be reviewed, such as breakdowns, newly discovered causes, overlooked factors, unanswered questions, and other plans for the future.
- To present case studies of improvements and abnormalities discovered, as well as improvement examples from other workshops.
- To list the number and types of abnormalities you have found.

When most, if not all the preceding information is documented, activity boards can describe your team activities in detail. This helps team members keep in mind what problems remain. It also helps promote cooperation and friendly competition among different groups.

Meetings and Reports

Enthusiastic meetings attended by all group members are essential for productive team activities. During autonomous maintenance implementation, it is especially important that all members participate and pool their collective wisdom to work out plans. Team activities will not work unless everyone feels part of the group. Discussions should not be dominated by one or two people; rather they should always be conducted with a team spirit.

Your group is responsible for ensuring that a report is filled out to record the meeting's contents, the conclusions reached, the next meeting date, and other facts. This should be sent to the area manager (see Figure 8).

These reports help keep others informed of what the teams are trying to achieve and the problems encountered on the way. In addition, the reports enable group members to receive advice and other feedback from management teams. It does not matter how complimentary, critical, or detailed the comments are; the important thing is that the group receives feedback from managers responsible for supporting their activities.

Group Activities Report

Date issued: October 18

Group name: Helical Circle

Department: Third Tool Dept., Group 1, Section 5

Group leader: Davies Secretary: Comarella

Theme: Preparing for the Step 2 audit

Participants: Comarella, Skinner, Labelle, Burns, Valdez

Absentees: None

Activities

- Shop-floor work:
- Meeting: Oct. 15, 3:50 to 5:00 pm
- Classroom activities:
- Total work-hours: (1.1 hours) x (5 persons) = 5.5 work-hours

No.	Item	Description of action or plan	Start date	Done by
1.	Audit on Step 2 scheduled for Oct. 22. Discussed what needs to be done before then.	1) Review initial cleaning (both day & night shifts must devote 15 minutes after the break period to thorough initial cleaning.)	Oct. 8	All
		2) Make list of most important items for initial cleaning.	Oct. 8	Labelle
		3) Post descriptions of oil leakage countermeasures (tagging, etc.) on activities board.	Oct. 20	Valdez
2.	Installed cover to localize debris on MC unit, but the unit is not being used now, so the effect of the cover has not been tested yet.	1) Use D9 MC to make model for testing effectiveness of improvement.	Oct. 20	Skinner, Burns
3.	Still no progress on improvement of intake port (response to source of problem).	1) Experiment with DP MC, keep statistics on grinder powder.	Oct. 20	Skinner, Burns, Comarella

Comment by Dept. Mgr.:	Comment by Dept. TPM Office:	Comment by Factory Group Leader:
Group should read up on Step 3	Should hold meetings more often	Use the PDCA cycle

Figure 8. Group Activities Report

6 Companywide Cooperation in TPM

TPM helps improve a company's competitiveness by eliminating the waste of equipment-related losses and improving quality. As was mentioned earlier, this is accomplished through our efforts to maximize two important resources — people and equipment. We cannot raise the level of equipment performance significantly or permanently without also changing people's attitudes and raising skill levels.

We have looked at how this can happen in the production department through an autonomous maintenance program. Ultimately, however, these changes must occur companywide, because isolated results — no matter how positive — are hard to maintain when the rest of the environment is not committed. A work environment that supports these changes must be created and sustained and top management must take the lead in making it happen.

PREPARING FOR TPM

Careful, thorough planning and preparation are the keys to successful companywide implementation. The preparation stage can last from three to six months, during which top management takes steps to create an environment that will support the introduction of TPM activities like autonomous maintenance and equipment improvement activities. Often management will announce its decision to introduce TPM and launch a formal education program designed to inform everyone in the organization about TPM activities and their benefits and how each department and individual can contribute. This step is important. It helps overcome the skepticism and resistance that often come up when management adopts new strategies that are not well understood by the people who are expected to support and implement them.

Assuring Organizational Support for TPM

A second important preparatory step is establishing a TPM organization that will promote and sustain TPM activities once they are begun. Every new project needs an organizational safety net to help people weather the occasional discontent and doubts that blow up during the early stages. This organization is team-based, because TPM activity brings diverse groups together and thrives on teamwork. Overlapping groups from top management to the shop-floor plan, coordinate, support, and implement TPM activities. Team leaders participate as members in the teams at the next level above their own and serve as a link between the levels. This structure promotes communication and helps guarantee that people at every level are working toward common goals (see Figure 9).

Planning for TPM Implementation

The first task of the TPM management team is to establish basic TPM policies and concrete, quantifiable goals. This will help teams at the middle-management and shop-floor level define and continuously refine their own concrete targets. To ensure that these goals are achieved, the management team also outlines a detailed master plan (three to four years) for introducing and implementing TPM. This plan identifies on the front side what resources will be needed when for training and education, equipment restoration and improvements, maintenance management systems, new technologies, and so on.

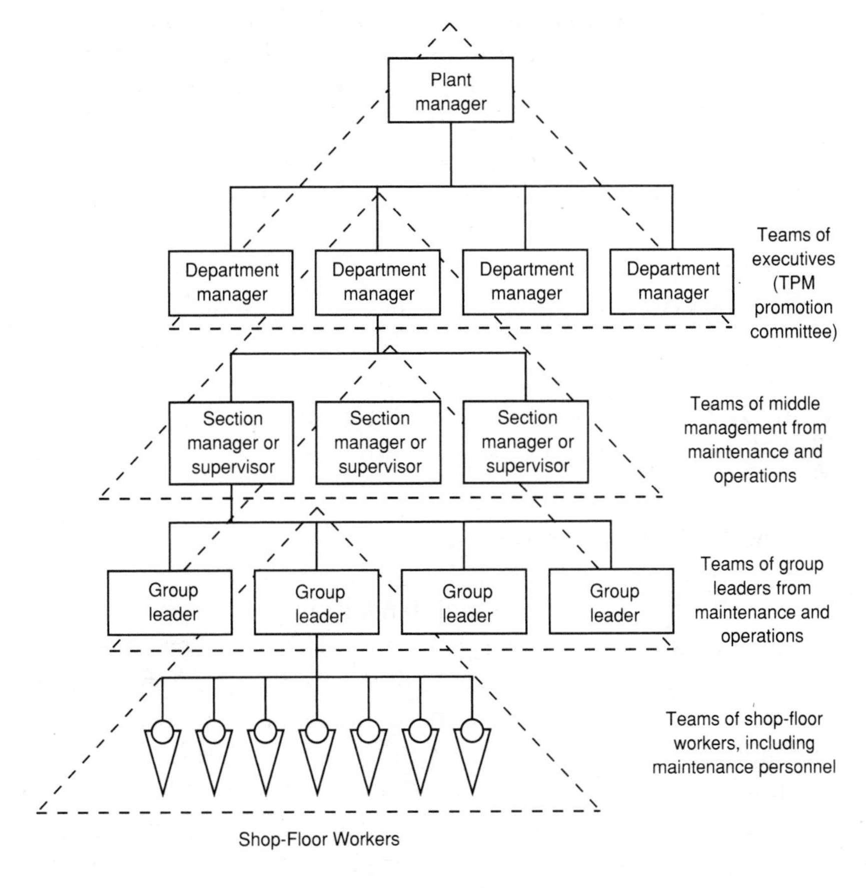

Figure 9. TPM Overlapping Team Organization

IMPLEMENTING TPM

When planning, preliminary education, and organizational development efforts are complete, actual implementation can begin in earnest. Autonomous maintenance by operators and equipment improvement activities by engineering/maintenance/production project teams often begin together. Both are aimed at reducing chronic losses and creating an environment that keeps equipment performing at the highest levels. Operators' routine cleaning and inspection help stabilize conditions and stop accelerated deterioration. They also share their experience with specific equipment problems and ideas for equipment improvements with equipment project teams. In turn, project teams' improvement results are often reflected in new autonomous maintenance inspection items or operating procedures.

Chapters 4 and 6 described equipment improvement activities carried out by project teams and autonomous maintenance performed by the production department. There are more specialized activities for reducing breakdowns and defects, however, that must be performed by professionals in the maintenance and engineering departments.

Planned Maintenance Activities

Planned or preventive maintenance improvement is led by the maintenance department, but carefully coordinated with ongoing autonomous maintenance and equipment improvement activities. In the early stages, the maintenance department will probably be kept very busy correcting problems brought to their attention by operators and project teams. This will be in addition to their daily work repairing accidental breakdowns and carrying out preventive maintenance tasks. Moreover, they will take on a new role as teachers and guides, providing training, advice, and equipment information to the teams. The volume of their work will diminish once again as autonomous maintenance becomes part of the operators' routine, and as breakdowns and emergency repair work decrease due to more thorough preventive care and improvements in equipment reliability. At that point, the maintenance department is freer to focus on improving its own organization and methods.

Specialized Work of the Maintenance Department

Some activities are beyond the scope of autonomous maintenance by production workers; for example:

- Tasks requiring special skills
- Overhaul repair in which deterioration is not visible from the outside
- Inspection and repairs that require significant disassembly
- Tasks requiring special measuring techniques and tools
- Tasks posing substantial safety risks

The maintenance department should handle all these tasks and also periodically check to ensure that nothing has been overlooked in the inspection work done by the production department. In too many factories, maintenance staff are too busy responding to sporadic breakdowns to carry out periodic and overhaul (disassembly) inspections; in the absence of the necessary preventive measures, breakdowns continue. The greater the number of breakdowns, the less time there is for the maintenance department to carry out planned maintenance. This whole situation works against the possibility of ever reducing the number of breakdowns.

Obviously, neither autonomous nor planned maintenance, by itself, can succeed in reducing breakdowns and defects. Instead, these activities must work together like the two axles of an automobile, and each situation must be studied to see how one can support the other. The maintenance department can help by giving advice for autonomous maintenance and promptly responding to the production department's request to treat abnormalities. As operators learn more about their equipment through general inspections, it will be useful to review the division of labor and determine whether the production department is ready to perform tasks that until then had been the responsibility of the maintenance department. For each type of equipment the maintenance professionals should determine how much cleaning, inspection, repair, and parts replacement should be done by the production department and how much requires their more specialized skills.

As the maintenance department gradually passes more work to production, their extra work-hours can be directed toward improvement-related maintenance activities.

A Quick Response System

The maintenance department must respond quickly to abnormalities discovered through autonomous maintenance. Equipment operators deserve an explanation when problems cannot be fixed immediately due to technical difficulties or cost considerations. When no obstacles exist, the maintenance department should fix problems right away and take the opportunity to explain the response being taken.

When responding to breakdowns, the maintenance people need to work closely with production personnel, not only by taking emergency measures but also by checking for prior symptoms, searching for causes, checking results of the measures, determining whether the same problem is likely to occur in similar equipment, and planning preventions. If possible, they should look into these things the same day the breakdown occurs.

This cooperation between the maintenance and production departments is important for the following reasons:

1. It helps operators take responsibility for breakdowns in their own equipment, shows them what symptoms show up prior to certain breakdowns, teaches them how to respond, and gains their cooperation in taking preventive measures against future breakdowns.
2. It also helps maintenance people improve their own skills by providing opportunities to check the quality of their work, better understand the causes of breakdowns, and learn how to plan preventive measures.

ACTIVITIES AIMED AT EARLY DISCOVERY OF ABNORMALITIES

There are two strategies the maintenance department pursues to discover and prevent abnormal conditions before they cause failures or other equipment losses: time-based or scheduled maintenance and condition-based or predictive maintenance.

Scheduled maintenance involves periodic inspection to detect and correct deterioration. And, because there are so many inspection points for each piece of equipment, we use a maintenance calendar — often computerized — to tell us when to conduct inspections, overhauls, oil changes, parts replacement, and so on, for each piece of equipment. Your autonomous inspection activities are included on this calendar. And, like your activities, each task on the calendar is subject to inspection and disassembly standards to ensure that they are always carried out correctly.

Predictive maintenance uses diagnostic devices to measure equipment deterioration or to detect the subtle symptoms of abnormal conditions. These tools permit more accurate prediction of where and when abnormalities are likely to occur without disassembling the equipment by measuring such things as vibration, heat, or the chemical composition of lubricating oil. They also allow us to avoid unnecessary repairs and to check the quality of necessary ones. Often these measurements can be taken by operators using relatively simple tools.

Activities to Prevent Chronic Failures

Production and maintenance personnel cooperate to prevent the recurrence of failures in several ways. For example, teams of production, maintenance and engineering people study the causes of chronic problems and develop preventive measures that might include adding specific cleaning or inspection items to maintenance procedures, changing the setup or operating procedures, or in some cases modifying the equipment to address a specific weakness in the design. Operators' improvements to control the spread of contamination or to make equipment easier to inspect and service are examples of such modifications. Other improvements that extend life include

making the equipment more durable and reducing the complexity of its mechanisms and systems.

Collecting and Using Maintenance-related Information

An information system is fundamental for good maintenance because it provides valuable data on equipment improvements, breakdowns, and breakdown analysis, and so on. These systems are often computerized, but not always easy to use. Often, however, production workers can help maintenance gather this important data. This becomes possible when production and maintenance people agree on what kinds of information are useful and why, and what form the data should be in. Then they can design a system that makes it easy for operators to contribute what they know.

MP Design

One of the goals of TPM is to develop maintenance-free equipment. One way to do this is to make improvements at the earliest possible stage; in other words, when the equipment is designed. TPM includes activities aimed at preventing breakdowns and defects in newly installed equipment by applying preventive maintenance principles during the design process. In other words, MP design includes discovering weak points in currently used equipment and giving feedback data to the design engineers.

The search for weak points in equipment can be carried out from the following perspectives:

Facilitating Autonomous Maintenance

Some of the most valuable information designers need comes from production workers who must deal every day with the limitations of equipment that was not designed with maintainability or operability in mind. For example:

- Can cleaning and inspection be easier?
- Can lubricating be centralized so lubricant is supplied at just one or two inlets per equipment unit?
- Can the piping layout be improved by changing the location of FRLs or solenoids, for example?
- Can the length and layout of wires be improved?
- Is debris scattered around?

Increasing Ease of Operation

- Can equipment be more resistant to operator errors, such as by changing the positions of switches and the layout of buttons on control panels?
- Can changeover procedure be simplified?
- Can standards be clarified to facilitate adjustments, or can measurement methods be made easier?

Improving Quality

- Have the precision settings and methods been determined (what to measure, how to measure it, limit values, etc.)?
- Is precision easy to measure?
- Is diagnostic equipment easy to set up? Does it have visual displays?

Improving Maintainability

- Have equipment life data been collected, and is work in progress to extend equipment life?
- Can parts replacement be simplified?
- Can replacements be done in preassembled units?
- Are breakdown analysis and repair measures applied to prevention measures in similar equipment?

- Are self-diagnostic functions built into the equipment?
- Can oil supply and oil changing be simplified?

Safety

- Are interlocking methods safe?
- Are there safety fences around hazardous equipment?

The factory-floor people and maintenance staff should give feedback on these matters to the design department so it can incorporate maintenance-preventive improvements into the equipment. Naturally some improvements cost more than others, and equipment design engineers must weigh the cost of each suggestion against the estimated savings.

EARLY EQUIPMENT MANAGEMENT

When new equipment is installed, problems often show up during test-running, commissioning, and start-up, even though design, fabrication, and installation appear to have gone smoothly. During this period, production and maintenance engineering, people work hard to eliminate "bugs" in the new equipment. They must often make many improvements before normal operation can begin, to correct problems caused by such problems as 1) poor selection of materials at the design stage, 2) errors occurring during fabrication of the equipment, or 3) installation errors. The delays caused by such problems are very costly. Even then, the repairs, inspections, and adjustments needed during the start-up period, and the initial lubrication and cleaning needed to prevent deterioration and breakdowns are often so difficult to carry out that supervising engineers become thoroughly discouraged. As a result, inspection, lubrication, and cleaning may be neglected, which needlessly prolongs equipment downtime for even minor breakdowns.

Many of these troubles can be avoided when the appropriate processing and operating conditions are built into the equipment through the application of MP design principles. Early equipment management also minimizes these errors or omissions and the delays they cause by identifying or predicting them at the stage in which they occur and taking action at that time to prevent them. The key strategy is simple — the same one operators apply in improving their own cleaning and inspection procedures — thoroughly listing all abnormal conditions and systematically addressing each one.

Generally, people notice very few problems at the design stage, but the cost of correcting them later is considerably higher. Virtually 95 percent of LCC (life-cycle cost) is determined at the design stage. Certainly, maintenance and energy costs of operation are determined by the equipment's original design. Efforts to reduce LCC after the design stage will affect only 5 percent of the overall figure.

In many cases, unfortunately, there is poor communication between the equipment design department, the production department, and the maintenance department. This makes it difficult to use the information obtained from routine PM activities to design better equipment. Maintenance engineers do not share data that could be relevant at the design and fabrication stages; and design engineers do not standardize general technical data or use the maintenance data they receive. When maintenance and design engineers cooperate to close the gap between maintenance and design technology, much waste can be avoided.

MP design and early equipment management both attempt to reduce the cost of human error and inadequate planning. MP design reduces the cost of the normal operating life of equipment, and early equipment management reduces the cost of the early failure period. By documenting well their observations and ideas related to equipment, production and maintenance personnel play an important part in reducing these costs.

This chapter has shown just how broad the scope of equipment management should be, and how careful planning is important to prevent costly oversights and omissions. To smooth the ride on the road to "zero breakdowns" and "zero defects," autonomous maintenance and specialized maintenance activities must be closely coordinated with the work of the engineering and design division.

About the Authors

Productivity Press has been the leading American source of books and media on TPM since 1988. Titles include *Introduction to TPM, TPM Development Program, Training for TPM, TPM for Workshop Leaders, Equipment Planning for TPM, Eliminating Minor Stoppages on Automated Lines,* and *Quality Maintenance,* as well as a video, *Total Productive Maintenance.*

Productivity Press is the publishing arm of Productivity, Inc., a full-service informational network dedicated to helping industry achieve continuous improvement in quality, productivity, and overall competitiveness. For more than a decade, the company has served as one of America's chief information links to production and management advances from around the world. Other services include conferences, seminars, in-house training and consulting, industrial study missions, and membership in the American Institute for Total Productive Maintenance, the Total Employee Involvement Institute, and the Process Management Institute.

Kunio Shirose is the executive vice president of the Japan Institute of Plant Maintenance (JIPM). He is JIPM's leading consultant and directs the TPM Consulting and Operations Division. He began his consulting career with the Japan Management Association and joined JIPM in 1984, shortly after it was established. He is also a member of the PM Prize committee.

Mr. Shirose has advised many companies in Japan and the United States. His consulting work focuses on improving quality control in human-machine systems and increasing equipment efficiency. Some of his many consulting clients include Daihatsu, Nachi-Fujikoshi, Yamagata NEC, Topy, Nissan, Matsushita Electric Works, Kansai NEC in Japan, and Ford Motor Company in the United States. His specialty is comprehensive TPM procedures and operations.

Mr. Shirose is the author of *TPM for Workshop Leaders* (Productivity Press, 1992) and coauthor of a number of other books on TPM.

ALSO FROM PRODUCTIVITY PRESS

Productivity Press publishes and distributes materials on continuous improvement in productivity, quality, and the creative involvement of all employees. Many of our products are direct source materials from Japan that have been translated into English for the first time and are available exclusively from Productivity. Supplemental products and services include membership groups, conferences, seminars, in-house training and consulting, audio-visual training programs, and industrial study missions. Call toll-free 1-800-394-6868 for our free catalog.

TPM for America

What It Is and Why You Need It

Herbert R. Steinbacher and Norma L. Steinbacher

As much as 15-40% of manufacturing costs are attributable to maintenance. With a fully implemented TPM program, your company can eradicate all but a fraction of these costs. Co-written by an American TPM practitioner and an experienced educator, this book gives a convincing account of why American companies must adopt TPM if we are to successfully compete in world markets. Includes examples from leading American companies showing how TPM has changed them.
ISBN 1-56327-044-7 / 169 pages / $19.95 / Order TPMAM-B208

Introduction to TPM

Total Productive Maintenance

Seiichi Nakajima

Total Productive Maintenance (TPM) combines preventive maintenance with Japanese concepts of total quality control (TQC) and total employee involvement (TEI). The result is a new system for equipment maintenance that optimizes effectiveness, eliminates breakdowns, and promotes autonomous operator maintenance through day-to-day activities. Here are the steps involved in TPM and case examples from top Japanese plants.
ISBN 0-915299-23-2 / 149 pages / $45.00 / Order ITPM-B208

TPM for Workshop Leaders

Kunio Shirose

A top TPM consultant in Japan, Kunio Shirose describes the problems that TPM group leaders are likely to experience and the improvements in quality and vast cost savings you should expect to achieve. In this non-technical overview of TPM, he incorporates cartoons and graphics to convey the hands-on leadership issues of TPM implementation. Case studies and realistic examples reinforce Shirose's ideas on training and managing equipment operators in the care of their equipment.
ISBN 0-915299-92-5 / 192 pages / $34.95 / Order TPMWSL-B208

Productivity Press, Inc., Dept. BK, P.O. Box 13390, Portland, OR 97213-0390
Telephone: 1-800-394-6868 Fax: 1-800-394-6286

Training for TPM

A Manufacturing Success Story

Nachi-Fujikoshi (ed.)

This training guide vividly demonstrates the value of TPM in the factory through a case study of Nachi-Fujikoshi, a Japanese manufacturer of industrial precision parts. The book details how the company trained managers and workers and shows the improvements they achieved in reducing breakdowns and defects and revitalizing the workforce. This study sets forth a model your company can follow in implementing TPM.
ISBN 0-915299-34-8 / 272 pages / $65.00 / Order CTPM-B208

Total Productive Maintenance

Maximizing Productivity and Quality (AV)

Japan Management Association (ed.)

Learn more about TPM with this accessible two-part audio visual program, which explains the rationale and basic principles of TPM to supervisors, group leaders, and workers. It explains five major developmental activities of TPM, includes a section on equipment improvement that focuses on eliminating chronic losses, and describes an analytical approach called PM Analysis to help solve problems that have complex and continuously changing causes. (Approximately 45 minutes long.)
ISBN 0-915299-49-6 / 2 videos / $799.00 / Order VTPM-B208

JOIN THE AMERICAN INSTITUTE FOR TOTAL PRODUCTIVE MAINTENANCE (AITPM)

The AITPM is your complete resource for learning how other companies are using TPM to eliminate machine breakdowns, increase equipment effectiveness, and improve equipment design so machines don't break down in the first place. Membership benefits include a monthly newsletter, networking opportunities, conferences, and special discounts on selected books and events. To sign up, or for more information about this item only, call 1-800-394-5772.

TO ORDER: Write, phone, or fax Productivity Press, Dept. BK, P.O. Box 13390, Portland, OR 97213, phone 1-800-394-6868, fax 1-800-394-6286. Send check or charge to your credit card (American Express, Visa, MasterCard accepted).

U.S. ORDERS: Add $5 shipping for first book, $2 each additional for UPS surface delivery. Add $5 for each AV program containing 1 or 2 tapes; add $12 for each AV program containing 3 ore more tapes. Include sales tax if you are ordering from MA, CT, or OH. We offer attractive quantity discounts for bulk purchases of individual titles; call for more information.

INTERNATIONAL ORDERS: Write, phone, or fax for quote and indicate shipping method desired. For international callers, telephone number is 503-235-0600 and fax number is 503-235-0909. Prepayment in U.S. dollars must accompany your order (checks must be drawn on U.S. banks). When quote is returned with payment, your order will be shipped promptly by the method requested.

NOTE: Prices are in U.S. dollars and are subject to change without notice.

Productivity Press, Inc., Dept. BK, P.O. Box 13390, Portland, OR 97213-0390
Telephone: 1-800-394-6868 Fax: 1-800-394-6286